Gerhard Beestermöller and David Little (Eds.)

Iraq

Gerhard Beestermöller and David Little (Eds.)

IRAQ

Threat and Response

LONDON AND NEW YORK

LIT

First published 2003 in Europe by LIT VERLAG
First published 2003 in North America by Transaction Publishers

Published 2017 by Routledge
2 Park Square, Milton Park, Abingdon, Oxon OX14 4RN
711 Third Avenue, New York, NY 10017, USA

First issued in paperback 2018

Routledge is an imprint of the Taylor & Francis Group, an informa business

Bibliographic information published by Die Deutsche Bibliothek
Die Deutsche Bibliothek lists this publication in the Deutsche Nationalbibliografie; detailed bibliographic data are available in the Internet at http://dnb.ddb.de.

Library of Congress Cataloging-in-Publication Data

Iraq : threat and response / Gerhard Beestermöller & David Little-editors.
p. cm.
Includes bibliographical references.
ISBN 0-7658-0207-4 (cloth : alk. paper)
1. Iraq–Military policy. 2. United States–Military policy. 3. Arms control–Verification–Iraq. 4. Security, International. 5. United States–Foreign relations–Iraq. 6. Iraq-Foreign relations–United States. 7. United States–Foreign relations–2001- 8. World politics–1995-2005. I. Beestermöller, Gerhard. II. Little, David, 1933-

DS79.755.I73 2003
327.730567'09'0511-dc21 2003044785

ISBN 13: 978-1-138-51107-1 (pbk)
ISBN 13: 978-0-7658-0207-1 (hbk)

Contents

Introduction

David Little

The publication of this collection of essays on the current crisis concerning Iraq will not be welcomed by the United States government. Although the authors – a group of German and American scholars, who are moral theologians, policy analysts, political scientists, and a Middle East historian – write from divergent backgrounds and perspectives, all finally concur, sometimes for different reasons, in rejecting the arguments of the Bush administration in favor of unilateral U.S. military action against Iraq.

That degree of unanimity was not entirely predictable when the authors joined a larger colloquium last June on the subject, "Iraq: Threat and Adequate Response", held at the Institute for Theology and Peace in Barsbuettel (near Hamburg), Germany. Yet, such agreement exemplifies an expanding consensus across national and disciplinary boundaries that although the Bush administration may deserve credit for some aspects of its Iraq policy, it is a long way from making a compelling case for its right to use force against Iraq as it alone sees fit. Indeed, judging from this collection, there is deepening apprehension that, on balance, the Bush policies may be leading the world in altogether the wrong direction.

A second reason these essays will not thrill the Bush administration and its supporters, is that they are uniformly free of the intemperate language and careless argumentation that characterizes some of the opposition to American policy inside and outside the United States, and is therefore easy to dismiss. Whether the authors address the moral, legal, political or historical dimensions of the Iraq problem, or whether they assess either the threat Saddam Hussein represents to his region and the world or the prospects for alternative strategies, the reasoning is generally well-informed, sensitive to complexity, and attentive to detail. The book will help to confirm and strengthen the growing 'thoughtful opposition' in the United States and abroad to the Bush policies, and as such deserves to be taken very seriously.

Considered as a whole, this book makes clear that a fundamental issue underlying the entire debate over Iraq policy is what we may call, "conflicting assessment of risk". The Bush administration and its defenders agree with opponents (such as are represented in this volume) that there are risks no matter what is done or not done in regard to Saddam Hussein. An attack against Iraq that succeeds in overthrowing him, particularly if it is not authorized by the U.N. Security Council, may be attended, the Bush administration concedes, by some unwelcome consequences. One is, as Secretary of Defense Donald Rumsfeld recently declared, "the loss of credibility and possible collapse" of the United Nations for having failed to do its duty, as he believes, in regard to disarming Saddam Hussein.[1] Another is substantial retaliation by Saddam, including the possible use of biological and/or chemical weapons, against Israel or other neighbors,[2] or a dramatic upsurge in anti-Americanism and in the number of terrorists and terrorist attacks against American personnel and assets around the world. Another troubling outcome would be instability in a post-Saddam Iraq, or evidence of reversion to militant Islam in the region.

Nevertheless, these negative possibilities are, from the point of view of the Bush administration, either outweighed by a set of greater risks likely to occur if Saddam is not forcibly deposed, or they turn out, on inspection, to be less probable than other, more favorable, results. The Bush administration has long since convinced itself that the greatest risk of all is a Saddam armed with the WMD he is (according to them) surely developing, including, especially, nuclear weapons. As such, he is seen as a catastrophic threat to the region, and to the United States and its allies, by means of nuclear intimidation, or through expanded encouragement and support of terrorism. If there is some cost to preempting such an intolerable state of affairs, then that is the price that both must and can be paid. So far as the administration is concerned, the benefits either outweigh the costs, or are by no means as exorbitant as critics allege.

---

[1] "U.S. Resists Call for Second UN Vote on Attacking Iraq" *New York Times* (Jan. 16, 2003), p. A10.

[2] "U.S. Sends 600 Troops and Antimissile Systems to Defend Israel If Iraq Attacks", ibid.

For example, while the Bush administration would, presumably, regard the downfall of the United Nations as unfortunate, they appear to have no doubt that worse things could happen. In what has been described as "the most sweeping shift in U.S. grand strategy since the beginning of the Cold War",[3] the administration has officially announced a policy of worldwide military predominance matched by a solemn declaration that whenever and wherever it so decides, the United States will determine – if necessary, in defiance of the international system of rules and regulations – what constitutes an imminent threat to U.S. national security and what shall be done about it.[4] Given this attitude, which is only somewhat less condescending than the administration's response to other international institutions and arrangements, such as the International Criminal Court, the Kyoto Treaty, the International Comprehensive Test Ban Treaty, etc., it is hardly surprising that the demise of the U.N., should there be a conflict with self-defined U.S. interests and ideals, is regarded as a fully acceptable risk.

As to the danger that Saddam will employ biological and/or chemical weapons against Israel or other neighbors, the U.S. government presumably believes the antimissile systems it can supply will provide adequate defense. For its part, Israel, considered an ardent ally, and likely in its eyes to gain from a forcibly reformed Iraq, is willing, apparently, to run whatever risks there are. Similarly, the U.S. evidently has faith in the large numbers

---

[3] John Lewis Gaddis, "A Grand Strategy of Transformation" *Foreign Policy* (Nov.-Dec., 2002), p. 50.

[4] "The National Security Strategy of the United States" released by the administration on September 17, 2002: "While the United States will constantly strive to enlist the support of the international community, *we will not hesitate to act alone, if necessary, to exercise our right of self-defense by acting pre-emptively* ... " (emphasis added). The language strongly suggests that, in the last analysis, it is exclusively and totally up to the U.S. to decide what constitutes national self-defense and when and where it may initiate a pre-emptive attack. Even Secretary Powell, usually pictured as the lone defender of multilateralism within the administration, is himself ultimately loyal to what might be called a doctrine of preemptory unilateralism. Commenting on a possible breach by Iraq of Security Council Resolution 1441, which recently authorized an intrusive weapons inspections regime in Iraq, Secretary Powell stated that whatever the Security Council may decide, the U.S. "will reserve our option of acting" and will "*not necessarily be bound by what the Security Council might decide on that point*" (Quoted in Michael J. Glennon, "How War Left the Law Behind," *New York Times* (Nov. 21, 2002), emphasis added).

of national guard it promises to assign to U.S. installations at home and abroad for protection against the possibility of increased terrorist attacks.

But more importantly, the administration appears in general to discount purported threats of an upsurge of anti-Americanism, or of social and political disruption in a post-Saddam Iraq, or of a rise of militant Islam and a new generation of terrorists throughout the Middle East. President Bush and many of his senior foreign-policy advisors, such as Deputy Secretary of Defense, Paul Wolfowitz, are very optimistic about the general outcome of an invasion of Iraq. That outlook is rooted in "an almost missionary sense" of America's role in the world, and it includes "a vision of Iraq not merely purged of cataclysmic weaponry, not merely disarmed, but ... a democratic cornerstone of an altogether new Middle East". Accordingly, a reconstructed Iraq is to be the first step in a general and inexorable democratic transformation of the region, one in which the United States will come to be widely acclaimed and revered, and no longer disparaged and attacked by terrorists and others. This "moralistic streak" and "generally sunny view of the world's possibilities" may account for the affinity between advisors like Wolfowitz and a president who is similarly inclined.[5]

It should be added that so far as President Bush, and most of his administration is concerned, the risk of trusting in, or tolerating much longer, inspections and other nonviolent or 'less-violent' means of disarmament, like sanctions, monitoring, or deterrence and containment, is much more dangerous than resorting to force because, in the opinion of the administration, Saddam cannot effectively be restrained by these means. For the Bush administration, the point is fast approaching where the benefits of invading Iraq far outweigh the costs.

Obviously, many critics of the Bush policies, such as those represented in this book, evaluate the risks of unilateral invasion quite differently, and some critics (possibly also represented here) go farther and adjudge the dangers of using force at all at this point, even with Security Council authorization, to be greater than the benefits. As we say, opinions concerning Iraq policy divide precisely over conflicting assessments of risk.

A key point of conflict between the Bush administration and several of the authors of this book concerns the importance of the existing international system, including the United Nations. While President Bush and his

[5] Bill Keller, "The Sunshine Warrior" *New York Times Magazine* (Sept. 22, 2002), p. 51.

associates are ready to sacrifice the system, should it fail to endorse the U.S. appraisal of Saddam and the response believed by the U.S. to be required, five authors, Dicke, Giessmann, Langan, Christiansen, and Beestermöller, stand in strong opposition, if in differing ways.

Professor Dicke vigorously commends the United Nations Charter as the indispensable basis for "peace through international law". The Charter must be regarded by the United States, as by all other states, as the supreme authority regarding the international use of force, and, in his view, that principle applies without qualification to the Iraq situation. This conclusion is shared by Professor Giessmann, and both authors agree that the Security Council incontestably retains final authority in regard to determining "the existence of any threat to the peace, [or] breach of the peace" (art. 39), as well as the limits and conditions of exercising "the right of individual or collective self-defense" (art. 51).

Both authors imply that unilateral U.S. military action against Iraq in disregard of explicit Security Council action would be manifestly illegal, and one may conclude that for both of them the prospect of subverting Security Council authority, and thereby risking the collapse of the United Nations, is, under present circumstances, an unthinkable price to pay. It is also fair to infer from passing comments made about the U.N. by Professors Langan and Beestermöller and by Fr. Christiansen that they, as well, would give priority to U.N. rulings in regard to Iraq, and would give more weight to international institutions in general, and be willing to risk more to maintain them, than would the Bush administration.

Where various authors divide among themselves is over the connection of the prevailing international *legal* system – focused in the U.N. Charter – and questions of moral justification, including 'just-war reasoning'. Dicke holds the view that "the Charter rules out any resort to just-war reasoning",[6] and he takes issue with Beestermöller, who argues that in administering and enforcing a desirable international system, states must commit themselves, in what he thinks is a Kantian spirit, to conform their behavior to universal moral standards of peace and justice. These standards would inform the traditional just-war categories for evaluating the use of force, and, in certain circumstances, would permit states to override Security Council rulings, should there be good reason to believe that a given ruling is "legal but not

[6] Dicke, p. 1 of his chapter.

moral". (Beestermöller *does not* believe the present Iraq situation warrants such action on the part of the U.S. because, according to him, the Bush administration has not convincingly demonstrated its conformity with universal standards.)

Dicke complains that such a position expressly undermines the legal authority of the U.N. Charter, and, by implication, weakens its efficacy as a bulwark of international peace and security. He also rejects the use of 'ethical considerations' to justify Charter non-compliance in the case of armed humanitarian intervention, as with the unauthorized NATO Kosovo campaign in 1999. For Dicke, there is no basis for such action in the Charter, which continues to prohibit using force against the "territorial integrity or political independence of any state" (art. 2.4), or "to intervene in matters that are essentially within the domestic jurisdiction of any state" (art. 2.7). He bolsters his position in favor of the priority of procedural justice by contending that three key founders of modern international law, Hobbes, Grotius, and Kant, all support his position.

The central issue here of the relation between law and morality is both complicated and important, and deserves comment. Dicke has a point. Legally understood, "decisions taken by the Security Council are mandatory", and as with any positive legal system, it is important that there be such a designated body. The idea that there might exist *legally permissible* rights to disregard and override a final authority is logically incoherent, since the authority, then, is no longer 'final'.

Among Dicke's revered triumverate, Hobbes would certainly concur, since for him right rests on might, and not the other way around. Whether Kant and Grotius would agree is more controversial. There is much in Kant's *Perpetual Peace* that sounds like Hobbes, for example, his disparagement of Hugo Grotius as one of 'Job's comforters'. For Kant, the use of the language of 'right' in the context of international conflict, such as is characteristic of Grotius and others, is completely inappropriate, since that kind of language "does not have the least legal force; nor can it have, since states as such are not subject to any common external coercion".[7] On the other hand, Kant could hardly have forsaken the idea that overarching

[7] Immanuel Kant, *Zum Ewigen Frieden* (Stuttgart: Philipp Reclam, 1965), p.32. I have slightly altered the translation of Carl J. Friedrich, ed., *The Philosophy of Kant* (New York: Modern Library, 1949), p. 443.

moral principles ultimately must guide the organization of law, as of all aspects of human life. Whether Kant is inconsistent here or just very subtle cannot be pursued further.

But, contrary to Professor Dicke's position, the case of Grotius, as Kant himself indicates, is altogether different. Grotius is elaborately explicit about the connection of moral language, including 'just-war reasoning' to international law. In fact, a strong argument can be (and has been) made that the U.N. Charter in several respects incorporates and institutionalizes just-war standards mediated by Grotius, such as 'legitimate authority' (art. 39), 'just cause' (arts. 39 and 51) and 'last resort' (chs. VI-VII). Moreover, it is Grotius' unmistakable position that "war ought not to be undertaken except for the enforcement of rights".[8] As such, he is "among the founding forefathers of our modern doctrine of human rights".[9]

The deep connection between morality and law in an influential case like Grotius illuminates why it is going too far, in one sense at least, to say that "the Charter rules out any resort to just war reasoning". If what we have said is correct, then the Charter itself is in part a product of just-war reasoning. What that means, among other things, is that the U.N. Charter itself is part of *an on-going moral discussion* about the proper limits of the international use of force. Therefore, moral reflection on the Charter is not only appropriate, but unavoidable.

The debate over armed humanitarian intervention, prompted by the Kosovo campaign, and mentioned by several authors, is a good illustration of this on-going discussion. The claim, developed by none other than U.N. Secretary-General Kofi Annan, that the Charter urgently needs reforming so as to modify the principle of state sovereignty, and permit intervention in face of egregious domestic human rights abuse, was simply a reassertion of part of the neglected moral legacy of the just-war tradition. Grotius, for one, articulated a ringing doctrine of humanitarian intervention aimed at correcting wrongdoing, and "protecting innocent persons". In the light of recent experience in the Balkans, Rwanda and elsewhere, it is unlikely that

[8] Hugo Grotius, *Prolegomena to the Law of War and Peace*, trans. by Francis W. Kelsey (New York: Liberal Arts Press, 1957), p. 18. See also, David Little, "Hugo Grotius and the Doctrine of Just War" in Norbert Brieskorn and Markus Riedenauer (eds.), *Suche nach Frieden: Politische Ethik in der Frühen Neuzeit I* (Stuttgart: Kohlhammer, 2000).

[9] Geoffrey Best, *War and Law Since 1945* (Oxford: Clarendon Press, 1994), p. 28.

such morally compelling claims can hereafter be ignored in discussions of Charter reform.

All of this is to say that the 'just-war reasoning', in which Beestermöller, Langan, and Christiansen all instructively engage, is fully pertinent and welcome in debates over the legitimacy of a prospective use of force against Iraq. While it is important to be clear over the distinction between legal and moral justification, there appears to be no good reason to disallow moral reflection as a critical and illuminating supplement to legal considerations. Such an approach is doubly welcome since George W. Bush, in contrast to his father during the Gulf War, has made so little attempt to defend his policies in just-war terms.

Of special interest is the fact that none of the authors in this volume who employ just-war language finds much basis for supporting the Bush policies. That is an important point on the merits, but also because of the contrast with the debates surrounding the Gulf War of 1991. In those days, it would have been unlikely to find unanimity among a given group of scholars on the moral questions concerning the use of force in response to Saddam Hussein's invasion of Kuwait. In the present setting, President Bush is not without his just-war defenders,[10] but there was, at the time of the Gulf War, considerably more division and debate among scholars and political observers.[11] The explanation is for this significant variation requires further reflection.

As to the merits of the case in just war terms, Langan remains unconvinced, in effect, of the administration's assessment of risk. He denies the administration has satisfactorily demonstrated that Saddam represents an imminent threat (just cause), or that the Iraq problem cannot be dealt with by measures short of full-scale invasion, such as inspections and containment (last resort), or that the consequences of invasion will not be worse than the benefits (proportionality and reasonable probability of success). On this last point, Langan perceptively challenges the dangers of a policy of preemptive unilateralism: "If we are not seen to be doing justice in a way [that] is plausible and intelligible to the rest of the world, we will be unable

[10] John F. Cullinan, "Preempt Iraq", *National Review On Line* (Dec. 16, 2002), <http://www.nationalreview.com>.

[11] Fr. John Langan, as simply one pertinent example, provided a much more divided assessment of the justice of the Gulf War than he has in respect to the present Bush policies.

to build a lasting structure of peace which will include reliable means of controlling the spread of the weapons of mass destruction and of ensuring they are not used."

Beestermöller implies a similar critique to the effect that having failed to demonstrate its wider commitment to the universal standards of justice and peace in supporting its Iraq policy, the United States must be seen to be promoting its own self-defined interests, and consequently to be undermining, rather than serving, the cause of universal peace. Christiansen provides a summary of the Vatican position on Iraq policy, which yields conclusions very similar to those of Langan and Beestermöller.

The essays of Professor Fürtig and of Messrs Cortright, Miller, and Lopez serve to support the positions of Langan, Beestermöller, and Christiansen. Fürtig provides a very comprehensive, detailed, and dispassionate assessment of the threat represented by Saddam Hussein in respect to his capabilities regarding biological, chemical, nuclear weapons and delivery systems, as well as any possible support for terrorism. Fürtig concludes that "there is not enough evidence of imminent danger of Saddam Hussein's WMD program to risk a military campaign that could cause more harm than good and instability for the peoples of the region as well as for the whole world than the *potential* threat the Iraqi Baath regime [represents] might do in the future".

The paper by Cortright, Millar, and Lopez makes two important contributions. First, without minimizing the risk represented by Saddam Hussein for regional and international security, the authors produce a plausible case for doubting both the gravity and the imminence of that threat. They argue that as the result of sanctions, inspections, and various forms of financial and military containment, "Iraq's ability to produce WMD and to deliver them have been curtailed". In just-war terms, the authors argue that the decisive point of last resort, when all less-violent means of resolving a conflict have been exhausted, has not yet been reached in the case of Iraq. The moral and policy implication is that before employing force, policy-makers are duty-bound to give less violent means a chance.

Second, the authors describe with considerable care the three forms of less violent means, and sketch out how they may be implemented. They are sanction reform, facilitation and expansion of inspectors, and 'enhanced constraint' by means of an improved border monitoring system.

Whether the proposals are, from a technical point of view, feasible and practicable, the proposals deserve to be considered and discussed. They represent an enormously valuable contribution to this volume.

Peace Through International Law and the Case of Iraq

Klaus Dicke

In 1985, Wilhelm G. Grewe ended a lecture on "Peace Through Law?" with the following "concluding formula":

"The prevention of war and of any use of armed force must remain the paramount aim of peace politics. 'Peace' should be understood as an international process free from any use of force and directed towards the prevention of the use of force, a process in which by way of agreement and compromise such conditions of the coexistence of states and nations are created which neither endanger their existence nor violate vital interests of one or more of them in such a way that they feel forced to resort to the use of force after peaceful remedies are exhausted. This process must be materialized by the establishment of an international order which grants security and stability without excluding the possibility of peaceful change."[1]

From the point of view of political ethics, it is this formula against which any policy against Iraq must be tested because it quite comprehensively concludes the very essence of modern international law as established in particular by the Charter of the United Nations. The Charter, in turn, has to be regarded as the legally binding constitution of world peace.[2] As the first chapter of the present paper argues, the very essence of modern international law as established by the Charter rules out any resort to just war

[1] Wilhelm G. Grewe, Friede durch Recht? Berlin/New York 1985, 30 (author's translation).

[2] Cf. Klaus Dicke, Die UN-Charta – Ausbau und ungenutzte Möglichkeiten, in: Nach Überwindung des Ost-West-Konflikts. Gedanken zur "Neuen Weltordnung", hrsg. von der Hanns-Seidel-Stiftung, München 1994, 48 – 75.

reasoning in whatever modernized or dynamic version it may come along.[3] The second chapter will turn to the case of Iraq from both a legal as well as from a political science point of view.

I. Peace Through (International) Law – the Concept

Although Grewe's fomula as quoted above represents the essence of the concept of "Peace through international law", this concept is far more complex than any brief analysis or even a formula can show.[4] The concept emerged centuries before the League of Nations Covenant, the 1928 Paris Treaty and, finally, the U.N. Charter established the modern international law of peace. It combines political experiences from five centuries, but only after modern wars of mass destruction and two world wars had "brought untold sorrow to mankind"[5] States were prepared to transform this concept into legal and political reality. The first effort to undertake such a transformation, the League of Nations, failed, and this is not the place to go into the details of whether it failed because of its legal deficits or because of a lack of political will to apply it and to comply with its leading norms. The second effort to realize peace through law, the United Nations, was more successfull at least in so far as it survived the cold war, further developed international law to a remarkable degree and produced innovative instruments to implement the concept of peace through law by use of the peacekeeping forces, the International Criminal Tribunal, sanctions regimes and many others.

The U.N.'s core political mechanism, the Security Council, however, which as a political organ was an innovation, too, needed nearly fifty years to become as effective as the Charter had provided for: after 249 vetos until 1989 and very few exceptional cases of agreed action – e.g. Korea, South-Rhodesia, and peace plans for Palestine and Namibia – it was SC Reso-

[3] For such a modernized and dynamic version see, Gerhard Beestermöller, Krieg gegen den Irak – Rückkehr in die Anarchie der Staatenwelt? Ein kritischer Kommentar aus der Perspektive einer Kriegsächtungsethik, Stuttgart 2002.

[4] Cf. Jost Delbrück/Klaus Dicke, Zur Konstitution des Friedens als Rechtsordnung, in: Uwe Nerlich/Trutz Rendtorff (eds.), Nukleare Abschreckung – Politische und ethische Interpretationen einer neuen Realität, Baden-Baden 1989, 797 – 818; Jost Delbrück, Die Konstitution des Friedens als Rechtsordnung, Berlin 1996.

[5] Charter of the United Nations, Preamble.

lution 660 of 1990 which opened a new chapter in the Security Council's history.[6] This resolution reacted on behalf of the international community on Iraq's annexation of Kuwait. 12 years and more than 750 mostly unanimously adopted resolutions later the case of Iraq puts the concept of peace through law to a serious test once more.

In order to reconstruct the Charter's concept of peace through international law, a fresh look into both its history of ideas and its legal history is called for to shed some light on the leading experiences and different elements which formed it.

1. Peace Through Law in the History of Ideas

The concept of peace through law can neither be reduced to one single source nor can it be derived from one single philosophical tradition. Rather, it combines at least three fundamental insights into the nature of public law and into the establishment of peace by law which can be associated with the writings of Hobbes, Grotius and Kant and their related efforts to draw the consequences from the wars of their times.[7]

The Hobbesian element in the concept of peace through law reads "auctoritas, non veritas facit legem".[8] The argument of this formula is a twofold one: first, it accomodates politics and public law to the basic fact of a plurality of conflicting interests and belief systems. I call this the argument for pluralism and freedom. Secondly, the formula concludes Hobbes' argument that political agreement on non-violent behaviour is a *sine qua non* of peaceful pluralism and that a political body must be established to secure this agreement – if necessary by police force. I call this the argument for a legal order.

[6] Michael Bothe, Militärische Gewalt als Instrument von Konfliktregelung: Versuch einer rechtlichen und politischen Ordnung zehn Jahre nach dem Ende des Ost-West-Konflikts, in: Sabine von Schorlemer (ed.), Praxis-Handbuch UNO. Die Vereinten Nationen im Lichte globaler Herausforderungen, Berlin 2003, 13–26 (17); Klaus Dicke, National Interest vs. The Interest of the International Community – A critical Review of Recent U.N. Security Council Practice, in: Jost Delbrück (ed.), New Trends in International Lawmaking – International 'Legislation' in the Public Interest, Berlin 1997, 145–169.

[7] Cf. Hugo Preuß, Nationaler Gegensatz und internationale Gemeinschaft, in: Id., Staat, Recht und Freiheit, Tübingen 1926, 345–361.

[8] Thomas Hobbes, Leviathan, ed. by Edwin Curley, Indianapolis-Cambridge 1994, XXVI.

Hobbes is often referred to as the father of legal positivism, and following Hobbes' reasoning law is defined by its authoritative enforcement capacity.[9] This reading of Hobbes, however, in some way confuses his epoch-making discovery with regard to peace. Hobbes of course identified public law with the enforcement capacity of the State; but the gist of his argument is that under the condition of religious, philosophical and ideological pluralism the authority of public law must be reduced to the one and single truth of security as a precondition of freedom. This argument goes along with a significant limitation of law itself: law governs human action and not the minds of people. In other words, Hobbes discovered that pluralistic societies in order to live in peace have to establish an artificial body of regulations to govern human conduct peacefully. This artificial body which is completely different from the lex humana of the natural law tradition provides for security which in turn is the necessary and sufficient precondition of peace.

In the history of peace ethics Thomas Hobbes marks a turning point. Hobbes introduced the pax civilis of public law into peace ethics as a normative level of its own: in his view peace rests upon political bodies providing for security. Security, however, cannot be derived theoretically but is the result of political agreement on laws which conduct human behaviour and which if necessary are to be enforced by public authority. It is remarkable enough that even critics of Hobbes like Rousseau and Kant recognized the fundamental truth of pax civilis: "A state of peace among men who live side by side is not the natural state (status naturalis), which is rather to be described as a state of war ... [T]he state of peace must be established."[10] Hobbes, beyond doubt, leaves many questions open. He did not provide for a sufficient answer to the question of how to secure the justice and moral legitimacy of artificially established laws, and he left those who asked for peace among States with the unsatisfying answer that the state among States is a state of nature, i.e. a state of war, and that peace among states could only be imagined as "the consequence of peacemaking action".[11] In search for a

---

[9] In an international law context, see Georg Dahm/Jost Delbrück/Rüdiger Wolfrum, Völkerrecht I/1, Berlin-New York 1989, 34 et seq.

[10] Immanuel Kant, Perpetual Peace. A Philosophical Essay. Translated with an Introduction and Notes by M. Campbell Smith, Bristol 1992, 117 et seq., with (Smith's) reference to Thomas Hobbes.

[11] Grewe (see note 1), 7.

lasting peace among States, however, Hobbes' theory of public law and his emphasis on the necessity of peacemaking action became the guiding path of 18th century peace ethics.

The second, the Grotian element of the concept of "Peace through Law" can be boiled down to a latin formula, too: "ubi societas, ibi ius". Grotius[12] realized that the source of legal regulations is not to be sought in laws of nature or principles of metaphysics – and in so far he concurred with Hobbes – but in the ordinary conduct of people forming a society. Interaction and cooperation in given societies develop clusters of behaviour and customs, and legal regulations are established where people follow such clusters and customs with a sense of obligation.

Hugo Grotius, born in 1583, five years earlier than Thomas Hobbes, today is recognized as the "father of international law". It is not by chance that the writings of Grotius which for almost 300 years were more or less forgotten were rediscovered at the end of the 19th century, i.e. at the starting point of modern international law. That can be explained by the relevance of Grotian thought for both the question of sources of international law and for a systematic evaluation of rules establishing the ius in bello. But a third reason seems plausible, too. In a certain way Grotius correctly understood the dialectics of cooperation: in realizing cooperation, people at the same time produce a work (opus), not as a result of their cooperation, but still in the process of enacting cooperation as such. When people engage in cooperation, they are guided by a sense of obligation against both: their partners as well as the common aim of cooperation. This obligation can be referred to as a disposition to law and to peace.[13]

This disposition, however, is ambigous in so far as cooperation does not necessarily follow peaceful aims or use peaceful means. States may cooperate in waging aggressive wars, and the mafia as well as Al-Qaeda is without any doubt an instance of effective cooperation. Nevertheless, in particular in international law language "cooperation" is positively connotated; this results from the experience that international cooperation since the last third of the nineteenth century has become the highway to international

[12] For an excellent introduction, see Hasso Hofmann, Hugo Grotius, in: Michael Stolleis (ed.), Staatsdenker in der frühen Neuzeit, München 1995, 52 – 77.

[13] For further development of this argument, see Klaus Dicke, Internationale Kooperation als politikwissenschaftliche Kategorie, in: Christiana Albertina 36 (1993) 5 – 16.

law, international order and the integration of the international community. Without the experience of effective cooperation during the First World War the Entente would not have been prepared to agree with Wilson's plans for a League of Nations, and significantly enough Wolfgang Friedman came up with the thesis that it was the emerging international law of cooperation which substantially changed the structure of international law in the 20th century.[14]

Given this ambiguity of cooperation, however, why then can cooperation be regarded as a disposition to law and to peace? Again, the history of the late 19th and early 20th century can provide an answer: the now emerging notion of "international organization" as well as the labelling of the League of Nations by German international lawyers as an "Arbeitsgemeinschaft" qualified international cooperation as cooperation under law. "International organization" meant both: the promotion of interstate cooperation in order to establish international legal regimes and at the same time its constitutionalization by ruling out non-peaceful aims and means, establishing procedures of dispute settlement, and providing for an institutional framework to govern international cooperation. International organization provided for peaceful change. Thus the process of "international organization"[15] picked up Grotius' idea of developing clusters of cooperation plus transforming the "sense of obligation" into legally binding rules.

Kant's legal philosophy, the third element of the concept of peace through law to be sketched out here, is founded on a theory of history. On ethical reasons Kant orgues that people and peoples must regard history in terms of justice and peaceful change.[16] In this perspective, history has to be viewed under the guiding principle of progressive constitutionalization and the progressive development of legal relations: between citizens in a given state, between states, and between states and mankind. Given the political circumstances of his lifetime, his conclusion for interstate relations

[14] Dicke (see note 13), with further references.

[15] For a broader analysis Klaus Dicke, Effizienz und Effektivität internationaler Organisationen, Berlin 1994, 45 ff.; Jost Delbrück, "Das Völkerrecht soll auf einen Föderalism freier Staaten gegründet sein". Kant und die Entwicklung internationaler Organisation, in: Klaus Dicke/Klaus-Michael Kodalle (eds.), Republik und Weltbürgerrecht, Weimar-Köln-Wien 1998, 181–213.

[16] Kant does not say: history is ... ; my interpretation follows Johannes Schwartländer, Der Mensch ist Person, Stuttgart 1968, 102 et seq.

postulates a foedus pacificum, a league of nations in order to eliminate not only a given war, as the peace treaties of the 18th century did, but war at all. And he was very outspoken with regard to the conditions of a lasting peace – far more outspoken, as any just war theory ever has been –, and one of those conditions reads: "No state shall violently interfere with the constitution and administration of another."[17]

Again, a closer look into Kant's writings is called for to get the gist of his argument. First, Kant acknowledged the ethical principle of compliance with law. He not only rejected a right to resistence but stated a moral obligation of citizens to act lawfully and legally. Second, although he developed liberal criteria to judge the legitimacy of laws he never justified any breach of existing laws in the name of legitimacy. Instead, he favored an approach of reform and evolution. In international relations he made this perspective explicit by making a difference between what is right in thesi and what States regard to be right in hypothesi and by substituting the idea of a world-republic by a "federation averting war" which can be expected (but cannot be forced) to ever extend over the world. The reason for Kant's reluctance, or even more: for his rejection to promote revolutionary visions of a world state is, third, a twofold one: a world-state cannot be controlled democratically, and it is not in agreement with popular sovereignty.

2. Peace through International Law – a Short Legal History

The Charter of the United Nations can be read as the second effort in history to translate the concept of peace through international law into an operating legal system. In doing so, it payed due regard to the elements as elaborated so far. The cornerstone of the Charter is Art. 2 para. 4 which prohibits any use of force and even any threat of the use of force in international relations. Prepared by the League of Nations Covenant and by the Paris Treaty of 1928 this norm definitely eliminated the hither to acknowledged right of States to wage war (liberum ius ad bellum) and, at the same time, it definitely delegitimized any just war theory.[18] In a legal sense, the Charter prohibits war at all. Only once the Charter language refers to war: the

[17] Kant (see note 10), 112.

[18] Jost Delbrück/Klaus Dicke, The Christian Peace Ethic and the Doctrine of Just War from the Point of View of International Law, in : German Yearbook of International Law 28 (1985) 194–208.

first preamble paragraph reads: "to save succeeding generations from the scourge of war". It would be contrary to the wording and to the spirit of the Charter to read Chapter VII which establishes rules governing "action with respect to threats to the peace, breaches of the peace, and acts of aggression" including self-defence under Art. 51 as a re-introduction of war into international law. The Charter instead definetely excludes war from international justice. By doing so, it followed Kant's conclusion: "There is no intelligible meaning in the idea of the law of nations as giving a right to make war; for that must be a right to decide what is just, not in accordance with universal, external laws limiting the freedom of each individual, but by means of one-sided maxims applied by force."[19]

The founding fathers of the U.N. Charter, however, were far from being idealists as, e.g., Woodrow Wilson to a certain degree was. They knew about the necessity to establish an enforcement machinery which would be more effective as the legal mechanism of the League of Nations Covenant was. The result was Chapter VII of the Charter which vested the Security Council with the power to decide politically on enforcement measures. The Security Council is the only organ which is legally entitled to determine whether or not a given situation is an agression, a breach of the peace or a threat to the peace, and the Council is empowered to take measures on behalf of all members of the United Nations after it took such a determination.

Legally, this mechanism has three consequences: first, military measures as decided by the Security Council have the legal status of police actions on behalf of the international community. Second, the decision on such police actions has been taken away from single states and was consigned to a multilateral organ acting on behalf of all members of the United Nations. And third, each and every unilateral use of force in international relations is prima facie illegal. These three consequences are even valid in cases in which states act under their right of self-defence according to Art. 51 of the Charter[20] or in those cases like the Kosovan one in which mil-

[19] Kant (see note 10), 135.

[20] "The use of force in self-defence is still considered to be illegal, in principle, but is justified because of the existing state of self-defence." Jost Delbrück, Structural Changes in the International System and its Legal Order: International Law in the era of Globalization, in: Swiss Review of International and European Law 1/2001, 1 – 36 (11); cf. Id., The Fight against Global Terrorism: Self-Defense or Collective Security as International

itary action was not authorized by the Security Council but, nevertheless, was held legal and mandatory by NATO.[21] The Charter allows self-defence under restricted preconditions only – "if an armed attack occurs", "until the Security Council has taken measures", under the obligation to report to the Council immediately, and under rules of proportionality. In cases of Security Council inaction like the Kosovan one a subsidiary power of a group of states or of a multilateral organ to recommend – not to order – enforcement measures can substitute the mechanism of Chapter VII. This exceptional substitute, however, neither justifies any unilateralism nor in particular something like an emerging institute of a "humanitarian intervention"[22] which once again would confuse the U.N. Charter's answer to the core question of quis iudicabit and open the door for unilateral decision and maxims by those states powerful enough to enforce their interests.

The picture of the U.N. Charter as the constitution of world peace would be incomplete without a few remarks on several ways the Charter has in store for peaceful change and the progressive development of international law.[23] Time and again the U.N. functioned as a forum of multilateral treaty-making. The General Assembly and the Secretaries-General are fora and instruments of prevention and mediation.[24] The International Court of Justice or the International Court of the Law of the Sea today legally resolve disputes which one century ago would have been regarded justae causae belli. And although the results of the U.N. in the field of disarmament and arms control are far from being satisfactory, Article 26 of the Charter recognizes the close interdependence of peace, human welfare and disarmament.

To conclude this chapter: Grewe's formula as quoted above is far from being idealistic but precisely describes what states after 1945 agreed to be

Police Action?, in: German Yearbook of International Law 44 (2001) 9–24 (13 et seq.).

[21] See Bothe, Militärische Gewalt (see note 6), 21, with further references); Klaus Dicke, Zur Situation der Vereinten Nationen nach dem Jugoslawien-Einsatz der NATO, in: Id./Helmut Hubel (eds.), Die Krise im Kosovo, Erfurt 1999, 73–78.

[22] Bothe (see note 6), 21; cf. Klaus Dicke, Interventionen zur Durchsetzung internationalen Ordnungsrechts: Konstitutives Element der neuen Weltordnung?, in: Yearbook of Politics 3 (1993) 259–283.

[23] For a further elaboration of the U.N.'s 'agenda for peace' Sven Bernhard Gareis/Johannes Varwick, Die Vereinten Nationen, Opladen 2002.

[24] Most recently, Manuel Fröhlich, Dag Hammarskjöld und die Vereinten Nationen. Die politische Ethik des U.N.O-Generalsekretärs, Paderborn u.a. 2002, 43 et seq., 253 et seq., 358 et seq.

the legal basis of their international conduct. It describes it in ethical terms, that is to say that it is conscious of the fact that there is necessarily a difference between ethics and ethos, between norm and reality, between legal prescriptions and day-to-day behaviour by people and by states. But Grewe's formula rests on the premise that compliance with existing international law is the basis for any development towards a further evolving constitutionalization of world politics.

Before I turn to the challenge which every legal system has to face: the challenge of non-compliance, some remarks on typical misperceptions of modern international law on the side of peace ethics seem to be in order.

3. Misperceptions of International Law

The revitalization of collective security after 1990 initiated a renaissance of peace ethics. In particular cases like Srebrenica, Somalia, Rwanda, Afghanistan and now Iraq gave rise to ethical considerations which before September 11, 2001 concentrated on the establishment of ethical criteria to judge so-called "humanitarian interventions".[25] Fortunate and important as the spread of related debates in democratic societies is, it is nevertheless problematic that those debates at least reveal a tendency to mistake, to delegitimize or even to ignore modern international law. The prevailing perception of modern international law in ethical debates seems to be that of a rudimentary, incomplete, uncertain, highly politicised and unsufficiently institutionalized set of rules at the disposal of States.

One of the most regrettable misperceptions of modern international law occurs whenever in cases of non-compliance with international legal norms ethical considerations to justify counter-measures make use of the language of war. Far from being a simple lapsus linguae such war language reveals first that the U.N. Charter is far from being a vivid "constitution" which is firmly established in the minds of its constituency. Second, the re-emergence of war language indicates that the current understanding of peace is less an understanding of a political task but rather continues the Hobbesian perception of peace as the absence of war. After September 11, 2001 war language seems to dominate international debates instead of asking what kind of peace strategies the situation calls for. Third, the language

[25] Most recently C. A. J. Coady, The Ethics of Armed Humanitarian Intervention, Washington 2002 (United States Institute of Peace, Peaceworks No. 45).

of war represents political traditions which follow a closed and dogmatic concept of group identity.[26] I will briefly discuss two misperceptions of international law, accordingly.

One of the most misunderstood norms of the U.N. Charter is Article 51 under which "nothing in the present Charter shall impair the inherent right of individual or collective self-defence if an armed attack occurs ... ".[27] Commentators in peace ethics read this proviso as a bulwark of State sovereignty against Article 2 para. 4, and one conclusion reads: "Those provisions of international law call for a political and legal clarification and for a global political security system which until now was established neither by decisions of the United Nations nor by other universally binding treaties." Moreover, the imbalance of Article 2 para. 4 and the right to self-defence is paralleled to a "significant oscillation" in Kant's "Perpetual Peace" between the positive idea of a world-republic and the negative substitute of a foedus pacificum.[28]

Although one has to concede that in particular the state practice of the United States time and again extended the "inherent right to self-defence" far beyond the original intent of the U.N. Charter and that insofar there indeed is need for "legal clarification"[29] the underlying interpretation of Article 51 is nevertheless misleading. First, it completely ignores that in legal terms a global political security system exists, and that at least since the end of the cold war it works. Chapter VII of the U.N. Charter establishes a system of collective security, and although nobody would deny that its efficiency and effectiveness could be improved the mechanism of Chapter VII goes far beyond the foedus pacificum Kant had in mind.[30] Second, the

[26] For a detailed analysis of war language and its implications, see Lothar Brock (forthcoming).

[27] Charter of the United Nations, Art. 51. See Albrecht Randelzhofer, Commentary, in: Bruno Simma et al. (eds.), The Charter of the United Nations. Commentary, Boston-Lancaster 2002, 788–806.

[28] Matthias Lutz-Bachmann, Weltweiter Frieden durch eine Weltrepublik? Probleme internationaler Friedenssicherung, in: Id./James Bohmann (eds.), Weltstaat oder Staatenwelt? Für und Wider die Idee einer Weltrepublik, Frankfurt a.M. 2002, 32–45 (34 et seq.).

[29] Cf. Christian Tomuschat, Der 11. September und seine rechtlichen Konsequenzen, in: EuGRZ 28 (2001) 435–540; Thomas Bruha/Matthias Bortfeld, Terrorismus und Selbstverteidigung. Voraussetzungen und Umfang erlaubter Selbstverteidigungsmaßnahmen nach den Anschlägen vom 11. September 2001, in: VN 49 (2001) 161–167.

[30] Cf. Klaus Dicke, Bedeutungswandel kollektiver Sicherheit in der neuen Weltpolitik?,

position as quoted above does not conceive Article 51 as an integral part of Chapter VII. By enshrining self-defence into Chapter VII, the Charter justifies self-defence as a means to enforce Article 2 para. 4, and only as a device to this end. Again: although the United States and some other states seem to regard Article 51 as a reservation of unlimited State sovereignty and as a residual element of the liberum ius ad bellum this is not in line with both the Charter's wording and spirit. Instead of defending this spirit and, by doing so, making the Charter a vivid constitution, ethical positions reading the Charter as quoted above open the door for the search for a "new world order" beyond U.N. Charter law which after all experience in history would be the order of the most powerful state.

The same is true for Beestermöller's ethical evaluation of policy options with regard to Iraq.[31] His perception of modern international law, of the "UN-Ordnung" in his language, is a second case in point. The main problem in Beestermöller's approach is the conditioning of compliance with international law and, by doing so, a significant weakening of its binding force. He not only refers to the notion of a (subjective) renunciation of force in international relations ("Gewaltverzicht") instead of taking the (objective) prohibition of the use of force as a starting point, but also his overall reasoning looks for ethical justification of violations of international law: In his view, the U.N. order does not rule out any chance that a state or a group of states "in full compliance with formal legality" corrupt this order and pervert it to the contrary or that they "in the name of law enforce illegitimate interests": "Is it really ruled out that the Security Council authorizes military action to restore world peace which in reality is directed to nothing else than the enforcement of power interests?"[32] To rule such corruption out, according to Beestermöller, states should apply just war criteria in order to ensure that their compliance would be suitable to promote a world legal order.

Beestermöller's teleology of world legal order seems to mistake the U.N.'s and in particular the Security Council's multilateral character. The

in: Dieter S. Lutz (ed.), Globalisierung und nationale Souveränität. Festschrift. Wilfried Röhrich, Baden-Baden 2000, 399–411; Id., Regionalkammern – ein alternatives Modell zur Reform des Sicherheitsrates der Vereinten Nationen?, in: Von Schorlemer (ed.) (see note 6), 695–705.

[31] Krieg gegen den Irak (see note 3); cf. his contribution to this volume.

[32] Beestermöller, Krieg gegen den Irak (see note 3), 37, 40, 36.

U.N. was established by states "to unite our strength to maintain international peace and security, and to ensure, by the acceptance of principles and the institution of methods, that armed force shall not be used, save in the common interest".[33] Different from the legal mechanism of the League of Nations, the Charter vests the political organ and mechanism, the Security Council under Chapter VII, with the power to politically decide on this "common interest". To take a decision a majority of the Council's 15 members is needed including the concurring votes of the permanent members. This procedure, of course, does not rule out any instrumentalization of the Council for national or other particular interests; and of course in particular the veto opens ways to misuse the Charter.[34] But, first, no political procedure is imaginable which would work without political interests; second, by requiring a majority vote Chapter VII to a large degree reduces the risk of biased decisions. This is not to say that the Council and its procedures are not in need of reform.[35] But for the time being decisions taken by the Council are mandatory. The recognition by the Council of "the threat Iraq's non-compliance with Council resolutions ... poses to international peace and security" was the basis of Resolution 1441 (2002) by which the Council after months of war language and sabre rattling brought international law back in. What is the resolution's contribution and perspective to "peace through international law"?

II. The Case of Iraq

Every legal system has to face the challenge of non-compliance and of persistent illegal behaviour. Without going into the details of Saddam Hussein's dictatorial regime, the cynical game he continously played with the international community and the situation within Iraq the present author takes it for granted that Saddam Hussein persistently violates international

[33] Charter of the United Nations, Preamble.

[34] Klaus Dicke, Institutionentheorie und internationale Beziehungen. Politiktheoretische Überlegungen aus Anlaß eines chinesischen Vetos, in: Othmar Nikola Haberl/Tobias Korenke (eds.), Politische Deutungskulturen. Festschrift für Karl Rohe, Baden-Baden 1999, 297–308.

[35] Cf. my most recent proposals in "Regionalkammern – ein alternatives Modell zur Reform des Sicherheitsrates der Vereinten Nationen?", in: Von Schorlemer (see note 6), 695–705.

law and that his regime remains to be a threat to the peace. "Saddam Hussein must be dealt with"[36] – this maxime was put forth by the U.S. government, by public debate both within and beyond the U.S. and resulted in a diplomatic process which on November 8, 2002 led to the unanimous adoption by the U.N. Security Council of Resolution 1441. It is not an exaggeration to praise this resolution as a victory of peace and a strengthening of the Security Council.[37] Before this resolution will be interpreted along the lines of the concept of peace through international law as eleborated above some preliminary remarks on the question of how to deal with Saddam Hussein and related policy debates are called for.

In particular in the U.S. there were strong calls for unilateral action by the U.S. As Kenneth M. Pollack put it: "The United States should invade Iraq, eliminate the present regime, and pave the way for a successor prepared to abide by its international commitments and live in peace with its neighbors."[38] This recommendation for unilateral action by the U.S. is not in accordance with international law because, first, the Charter of the U.N. confers the primary responsibility for the maintenance of international peace and security to the Security Council (Art. 24). As shown above, any unilateral action beyond the narrowly limited right to self-defence in case an armed attack occurs is thus ruled out. Second, the Security Council was seized with the matter: Starting with Resolution 660 of 1990, it not only constantly dealt with the situation in Iraq but took a series of measures. By Resolution 661 of August 1990 it posed sanctions upon Iraq and established a committee to administer the sanctions regime. Under Article 48 of the Charter, it authorized a coalition of states to restore peace and security in the region.[39] After "desert storm" it established an arms control regime, a regime of minority protection, enlarged the sanctions regime and authorized their enforcement, accordingly. In 1996 the Council introduced the "food-for-oil" programme, and by Resolution 1409 (2002) it again changed

[36] Kenneth M. Pollack, Next Stop Baghdad?, in: Foreign Affairs, March/April 2002, 32–47 (42).

[37] Christian Tomuschat, Der Sicherheitsrat ist gestärkt, in: Frankfurter Allgemeine Zeitung, November 11, 2002.

[38] Pollack (see note 36), 33.

[39] SC Res. 678 of November 29, 1990. See Ursula Heinz/Christiane Philipp/Rüdiger Wolfrum, Zweiter Golfkrieg: Anwendungsfall von Kapitel VII der UN-Charta?, in: Vereinte Nationen 4/1991, 121 – 128; Bothe (see note 6) with further references.

the sanctions system by starting to introduce elements of so-called "smart sanctions" and decided "to remain seized of the matter".

Policy debates before and after September 11, however, raised serious doubts on both the effectiveness and even the legality of the sanctions regime and on the efficiency of the Security Council's policy towards Iraq. The sanctions as imposed on Iraq were criticized by, i.a., Hans Count Sponeck, Dennis Halliday and some legal scholars. They hold that insofar the sanctions dramatically reduced food, health, water supply, shelter, and life expectancy among Iraqi civilian population they constitute violations of international humanitarian law of armed conflicts.[40] The issue was picked up by the General Assembly in general terms when its Millenium Declaration held that sanctions on a regular basis should be monitored and that negative consequences for civilians must be reduced to a minimum.[41] In a certain way, SC Resolution 1409 reacted to this move by the Assembly.

Since in December 1998 Saddam Hussein refused further cooperation with the U.N., neither the sanctions nor the arms control regimes effectively prevented Saddam's regime to develop and/or dislocate arms of mass destruction. This was the center of the public debate on how to deal with Saddam Hussein in particular in the United States[42], to a much lesser degree in Europe. After the U.S. government started with preparations to act on Iraq, if necessary unilaterally, the American position met growing criticism and resistance both domestically and abroad. Chancellor Schröder announced that Germany would not participate in any military action against Iraq even if mandated by the Security Council which, nota bene, was not in accordance with Germany's obligations resulting from membership in the U.N. Among the permanent members of the Security Council in par-

---

[40] Hans-Christof von Sponeck, Politisch wirkungslos und menschlich eine Katastrophe. Elf Jahre Wirtschaftssanktionen gegen den Irak, in: Blätter für deutsche und internationale Politik 11/2001, 1353–1358; Michael Brzoska, Der Schatten Saddams. Die Vereinten Nationen auf der Suche nach zielgerichteten Sanktionen, in: Vereinte Nationen 2/2001, 56–60; David Cortright/Alistair Millar/George A. Lopez, Sanctions, Inspections, and Containment. Viable Policy Options in Iraq, Goshen/Washington 2002 (Joan B. Kroc Institute for International Peace Studies, Policy Brief F3).

[41] GA res. 55/2 of September 8, 2000, para. 9.

[42] Karl-Heinz Kamp, Ein Militärschlag gegen den Irak? Die Argumente der Vereinigten Staaten, Konrad-Adenauer-Stiftung Nr. 71/2002; Ferhad Ibrahim, Irak und Iran in der Phase II des amerikanischen Krieges gegen den Terror, in: Aus Politik und Zeitgeschichte B 25/2002, 31–38.

ticular France and Russia opposed unilateral action and, instead, sought to reestablish arms control monitoring in Iraq. After a process of laborious and difficult negotiations the Council agreed on Resolution 1441 on November 8, 2002.

The unanimous vote by the members of the Security Council including the Arab State Syria is evidence enough that legally and politically this resolution can claim to represent the common interest of the international community.[43] It maintains world peace at least for a certain time and establishes a process which enables Iraq to regain cooperation and enables the international community to regain control. Remarkably enough, the basis on which this procedure is established is compliance by Iraq with international law.

In the preambula paragraphs, the resolution recalls ten previous resolutions by the Council and recognizes that Iraq's non-compliance constitutes a threat to international peace and security. It recalls that the 1991 ceasefire "was based on acceptance by Iraq of the provisions" of Resolution 687 and gives expression to its determination "to secure full compliance with its decisions". Insofar, the Council insists on Iraq's compliance. Politically, this is a precondition to uphold its authority and the Council underlines its determination to do so in para. 13, recalling "that the Council has repeatedly warned Iraq that it will face serious consequences as a result of its continued violations of its obligations". It decides, "to afford Iraq ... a final opportunity to comply" (para. 2).

While the decisions as quoted so far follow the "auctoritas non veritas facit legem"-maxim and establish a procedure which opens opportunities for peacemaking actions, the Council at the same time takes due regard to the Grotian principle of cooperation. It notes that a letter by the Minister for Foreign Affairs of Iraq addressed to the Secretary-General of the United Nations "is a necessary first step toward rectifying Iraq's continued failure to comply with relevant Council resolutions". Further, it refers to negotiations between Iraq and U.N.MOVIC and IAEA as "prerequisites for the resumption of inspections". By doing so the Council takes Iraq's preparedness and first steps to cooperate, as modest as they may appear, as steps toward political agreement and peace.

[43] My evaluation follows Tomuschat (see note 37).

And third, the Council follows the Kantian maxim of constitutionalization. It "decides to convene immediately upon receipt of a report" indicating omissions or any interference with inspection activities by Iraq "in order to consider the situation and the need for full compliance" (para. 12). This proviso clearly establishes that non-compliance by Iraq will course 'serious consequences' but is a case for further decisions by the Security Council rather than a case for unilateral or other unconstitutional action.

III. Conclusion

By adopting Resolution 1441 the Security Council in a certain way re-invented the concept of peace through international law and re-invented itself. The resolution gives evidence that collective security under the Charter is a viable means to establish peace even in situations of great tensions between States or between one single State and the international community. At the same time, it reveals that compliance with and obligation to international law is the most important precondition to maintain peace and security.

The Dubious Legitimacy of Preventive Military Action against Iraq

Hans J. Giessmann

U.S. President George W. Bush's statement in January 2002 on the "axis of evil" was crystal clear. The world would be much better off if "evil" regimes such as those of Iraq, Iran or North Korea disappeared. Although one might question the validity of such a determined list – especially if one asks what the difference in "evilness" really is between the three nations that are named and the others that have not been named by the U.S. President, such as the hardly democratic regimes of Musharraf in Pakistan or Sharon in Israel – George W. Bush has touched on a sensitive issue. He has correctly pointed the finger at an "open wound" of the present international system.

How can the international community, the United Nations in particular, tackle the issue of flagrant violations of human rights by nation states, both cross-border and domestically, and remain in accordance with the regulations of international law? How can universal legal standards be enforced within the borders of those nation states that, on the one hand, are sovereign members of the U.N. while, on the other, their ruling regimes tentatively breach fundamental norms of international law in areas under their control. How should the international community react to the apparent preparations for war by nation states, before any such war has broken out?

There is no doubt that the list of governments that do not act in full compliance with international law is much longer than the short list presented by the U.S. president. Even only in terms of the sheer dimension of noncompliance, the list could be expanded. Human Rights Watch, Amnesty International and other human rights organizations have frequently accused some dozens of states, including Western nations, of permanently ignoring specific standards of international law. Even the United States has been named in their reports because of the still exercised death penalty and, most recently in conjunction with the anti-terror combat where hundreds of people have been arrested in the U.S. without being given the right to a fair trial.

Who qualifies the offender, who qualifies the judge? Who constitutes the jury and who provides the required legitimacy of the sentence? The U.S. approach towards Iraq in particular has raised worldwide concerns about a selective application of standards of international law. What is condemned by the U.S. Administration in one case, is obviously tolerated or even neglected in others. This being said, however, the crucial issue which has to be handled properly is the legitimacy of "intrusive prevention".

The existing legal machinery provides only insufficient rules for proactive interventions in cases where the sovereignty of nation states may become challenged with the principles underlying the U.N. Charter of non-intervention into internal matters of nation states. Can the international community afford not to take action when there is an imminent military threat from a sovereign member of the U.N.? Can something be done before it turns into a war against other U.N. member states? What can be done proactively by the international community when a ruling regime massively violates international law and, at the same time, the U.N. proves to be too weak to enforce its standards of law within that country, either because the U.N. is unable or some of its members are unwilling to take action under Chapter VII of the Charter? The present U.S. administration tends to answer these questions by opting for unilateral or collective steps independent of a formal pre-advance legitimacy provided by the U.N. Security Council. "With the U.N. if possible and without the U.N. if necessary" this statement, made by the former U.S. Secretary of State, Madeleine Albright, summarizes quintessentially the present U.S. position on the relevance of the U.N. for decision-making in Washington. The adequacy of its reaction to the possible military threat from Iraq to its neighbours, and to international peace and security raises questions far beyond the single case of Iraq. Furthermore, the character and future of the post-world war order seems to be at stake.

This paper deals with the following three aspects:

- the legal legitimacy,

- political legitimacy,
- and military legitimacy of
- preventive military action against Iraq. Finally, some preliminary conclusions will be drawn as tentative alternatives, which should be considered.

I. The legal legitimacy

After the 2nd Gulf War, the United Nations Security Council adopted several resolutions, the most important of which, No. 687 (1991), called upon Iraq to renounce the possession of weapons of mass destruction (WMD), to destroy all biological and chemical stockpiles and production capacities, and to scrap longer-range delivery systems, namely missiles. It set up a supervising commission, the UNSCOM, which, notwithstanding the plenty of difficulties it was confronted with, did by-and-large a fairly successful job until 1998. From the very outset of the UNSCOM mission, however, questions arose – not only on the Iraqi side – about the scope, the intrusiveness and the duration of inspections, and, within this context, about the inter-relation between inspections, their results and the possible lifting of political and economic sanctions that had been imposed on Iraq after its aggression against Kuwait. The United States and the United Kingdom referred to the enforcement of Resolution 687 by setting up, on their own, so-called "non-flight-zones" in the north and in the south of Iraq although the U.N. Security Council had never before – nor thereafter – endorsed or vetoed this unilateral step formally. The dispute over the efficiency of inspections and the sanction regime escalated into a four-day military escalation in 1998, after which the inspectors were expelled from Iraq on the order of the Hussein Regime. In fact, the Baath-Regime did not permit the inspectors to return, so that since 1999 no more reliable information about Iraq's ongoing (or not ongoing) WMD-programmes has been available. Because the U.N. Security Council has not adopted a new resolution on Iraq since then, the U.N. can either act on the basis of the existing resolution or – since Iraq suspiciously has not acted in full compliance with its provisions – it can supplement or revise resolutions in order to enforce its decisions more effectively. It is solely subject to the U.N. Security Council, not to Iraq or any other single nation, to decide on this matter.

From a legal point of view, the unconditional return of inspectors is a legitimate demand by the U.N., as it is overdue. Baghdad's counter-claim, that a return of inspectors must be connected with a clear schedule for lifting the sanctions upon confirmation of non-evidence of WMD in Iraq, is understandable from a political and moral point of view, however, it is not justified from a legal point of view according to the given U.N. resolutions. It is exclusively up to the U.N. Security Council to decide whether or not successful inspections are a reason to lift the present sanction regime. This being said, however, it is clear that sanctions in general make sense only if they support specific action, i.e. if they provide incentives to change the behaviour of those who the sanctions are imposed on. Flat sanctions hardly provide incentives for changing the sanctioned state's behaviour if they don't create a link between behavioural change and the scope or duration of the sanctions imposed. And U.N. Secretary General, Kofi Annan, has repeatedly stressed the necessity of lifting sanctions once the inspections are concluded successfully.

While the legal legitimacy of ongoing inspections – unless the Security Council decides differently – is out of question, the same does not hold true with any proactive military pressure on Iraq without approval from the Security Council. That holds true all the more so for military intervention. The U.N. Charter defines two exceptions of the basic principle of the prohibition of force (as stated in Art. 2, Para. 4 of the Charter: "All Members shall refrain in their international relations from the threat or use of force against the territorial integrity or political independence of any state, or in any other manner inconsistent with the Purposes of the United Nations."):

- the individual or collective defence against a military attack in accordance to Article 51 of the Charter ("Nothing in the present Charter shall impair the inherent right of individual or collective self-defence if an armed attack occurs against a Member of the United Nations, until the Security Council has taken the measures necessary to maintain international peace and security. Measures taken by the Members in the exercise of this right to self-defence (...) shall not in any way affect the authority and responsibility of the Security Council under the present Charter to take at any time such action as it deems necessary in order to maintain or restore international peace and security").
- the exclusive authority of the U.N. Security Council to determine action

as laid out in Art. 41 and 42 under Chapter VII of the Charter if a threat to peace, the breach of peace or any act of aggression is imminent, in order to maintain or to restore international peace and security according to Art. 39 of the Charter.

It may be interesting to note here that the Bush administration, while putting more and more military pressure on Iraq after 9/11, had originally tried to provide legitimacy to this policy towards Iraq by referring to it as a combat against international terrorism. This combat, which has been increasingly questioned, as far as its scope and duration is concerned, was originally against Al-Qaeda and the Taliban-Regime of Afghanistan and was based on Article 51 of the Charter. From a legal point of view, however, the right to self-defence is neither unlimited nor unconditioned. As stated in Article 51 ("until"), the right to self-defence is limited in action until the Security Council decides to act on its own capacity. Furthermore, the right to self-defence is explicitly bound to an "armed attack".

Finally, the principle of proportionality also has to be applied to actions in order to maintain or restore international peace and security. In fact, Iraq did become an armed attacker by invading Kuwait and, after having done that, the U.N. Security Council took action by adopting several resolutions which aimed at restoring peace in the region. It can be concluded from this that no individual state can claim the right to self-defence independent of any action which the U.N. Security Council has already taken. The U.N. Security Council is the supreme and only authority to decide whether or not Iraq complies with its own decisions, and what sanctions are to be imposed on Iraq in the aftermath of a war. Since neither evidence has been found to prove any collaboration between the Baath-Regime and the terrorist network of Al-Qaeda, nor has Iraq attacked another state militarily, there is no legal legitimacy for any individual member state or any group of member states, under the provisions of the U.N. Charter, to declare war on Iraq under the given circumstances without endorsement by the U.N. Security Council.

II. The political legitimacy

The nature of the Baghdad regime is definitely not democratic. Under the rule of Saddam Hussein, thousands of people have been killed, human rights have been massively and flagrantly violated, ethnic minorities have been persecuted and the Iraqi army has used chemical weapons, both in the first Gulf War and against Iraqi citizens of Kurdish nationality in northern Iraq. However, there are other regimes in the world who have either acted or still act in a similar way without being punished or who do not even fear comparable pressure from the United States. Not even in the war against Milosevic, in the case of Bosnia and Croatia and the Serbian rule over Kosovo, did the U.S. Government claim the need for a change of regime in Belgrade as a *conditio sine qua non* for restoring peace in the region. Neither the cases of Chechnya nor of Tibet, Palestine or Kashmir have triggered off similar demands by the Bush administration. Russia, China, Israel, India and Pakistan, all these nations possess nuclear weapons. All these nations have seriously violated the rights of ethnic minorities. Some of them have even fought wars against their own people or neighbours.

It would be indeed too simple to conclude that there is much more risk involved in threatening or imposing sanctions on a nuclear state when compared with a threshold state such as Iraq. The status quo of the countries mentioned is also not completely comparable with the situation in Iraq – none of them has to comply with U.N. sanctions. But, at least in one respect, all these specific cases are well comparable with that of the Middle East, and that of Iraq in particular: Multilateral efforts to solve the conflicts behind these cases on the basis of international law and under the auspices of the international community have over time almost vanished into thin air.

But even the failure of political settlement does not provide enough reason to substitute policy with force. On the contrary, using force against a frozen conflict may rather complicate the environment for a sustained and peace-oriented conflict resolution. Even if force were applied effectively, and if it thereby led to a victory on the battlefield, it still may not eliminate the root causes of the conflict structures behind it. This is all the more so true for the Middle East, where not only one conflict but a cluster of mutually overlapping conflicts exist. The conflict between Israel and the Palestinians, and between Israel and its neighbours Syria and Lebanon, the growing spread of Islamic fundamentalism in Egypt, Saudi Arabia, Yemen and Iran, the situation of ethnic minorities in Turkey and Iraq, the fragile situation in Afghanistan, the delicate character of the regional anti-terror coalition, the sensibility of the Middle East oil market and the vulnerability of sea lanes, the growing gap between rich and poor, the breeding ground for cross border terrorism, the prevalent culture of violence in fragmented societies, the proliferation of small arms and WMD – all these issues contain sources of tremendous instability and escalation that might be even fuelled by any military action taken by the U.S., or as a U.S.-led European coalition against Iraq.

Whereas it is argued here that, both for reasons of equally handling comparable cases and of paying attention to the complexity of risks and problems involved, military action against Iraq cannot be legitimized. However, a response to the question on how to adequately deal with a regime that may obviously pose a threat to regional and global security, and how to achieve a change of behaviour in Iraq in order to prevent further destabilization is required. George W. Bush correctly pointed to the necessity that Iraq comply with all resolutions adopted by the U.N. However, U.N. Resolution No. 687 is not just about inspections.

It not only concerns the "disarmament of Iraq", as nowadays frequently suggested by the U.S. President. The resolution calls for "regional disarmament" and the establishment of a zone free of WMD warheads and delivery vehicles as a part of a more complex peace accord in the Middle East. This challenge cannot be attributed to Iraq only, but is as relevant for Israel, Jordan, Syria, Turkey, Saudi Arabia, the Emirates and, last but not least, the United States itself. And, so far, when compared with the inspections in Iraq, this task has not been fulfilled either. If a new resolution on Iraq is considered by the U.N. Security Council, and if such a resolution is based on those previously adopted by the Security Council, it cannot ignore the broader context in which confidence-building, arms control and disarmament is embedded.

It is, however, unclear whether or not a more intrusive resolution is necessary in general. Since the expulsion of inspectors from Iraq took place, no reliable information about the state- of-the-art of the Iraqi WMD-programme exists; assessments are contradictory. While the U.K. and U.S. intelligence sources have released non-specified allegations about ongoing programmes in Iraq, others have raised serious doubts as far as the justification of such allegations is concerned. The Vienna-based International Atom-Energy Organization (IAEO) has declared that it has no information about any ongoing nuclear projects in Iraq. The German expert Hans von Sponeck, after visiting several suspected installations in Iraq, denied the capability of Iraq to manufacture weapons of mass destruction. Similarly, the U.S. Inspector Scott Ritter argued that Iraq was hardly in a position to produce such weapons, including required long-range delivery systems. His view was shared by the designated head of the new U.N. Weapons Control Commission, the former IAEO Director, Hans Blix. Even the London-based International Institute for Strategic Studies (IISS), which recently released a study on the Iraqi threat that has been used by the Governments of the U.S. and the U.K. to justify their military planning, has clearly stated that Iraq would definitely need fissionable material from abroad to produce nuclear weapons. If these assessments hold true, two conclusions can be drawn.

First, a window of time does exist, within which inspections can contribute to clarify the picture and especially to withhold Iraq from acquiring components or technologies for any WMD projects.

Secondly, and just as important, an effective regime of non-proliferation has to be set up in the region. It should not be forgotten, that during the first Gulf War it was not just Russia, but also the West, the U.S., France, and Germany in particular, who supported the military build-up of Iraq and helped the regime to sustain the war. Instead of arming neighbouring countries – which by far cannot be considered stable democracies – with modern weapons and military technologies, the weapon supplier states should adopt a long-term perspective of military stability in the region that is based on arms control, disarmament and peaceful cooperation.

And finally, reliable criteria for imposing sanctions have to be discussed in the U.N. Security Council, mainly to oblige the players in the field to cope with the rules of U.N. Resolutions and decision-making. As far as Iraq is concerned, a lifting of sanctions is required as soon as possible or after a successful ending of inspections, for the most part, because it is the Iraqi people that have suffered the most from the international embargo and blockade policy.

III. The military legitimacy

The Bush administration, or to be more precise, the neo-conservative inner-circle within that administration, has vigorously neglected any alternative policy option, except for striking militarily against the regime in Baghdad either with or without an explicit mandate provided by the U.N. Security Council. This perspective is neither justified from a legal or political point of view. But it is also highly risky and ambiguous in a military sense. The only plausible legitimacy for military action is related to the simplistic fact that toppling the regime with "civilian means" from the outside is not very promising. If the Bush administration sticks to its declared goal to topple the Baath rule, it has no alternative other than to use military force independent of what the U.N. may discuss or decide. Yet, such military calculus contains enormous risks. Toppling a regime is a somewhat different target than changing its behaviour. Why should the Baath-Clique change its previous behaviour if the United States keeps to its non-negotiable claim of overthrowing the Regime in Baghdad, independent of what it is saying or doing. If the regime, however, is threatened with death, it will definitely change its attitude towards political concessions and will defend its position of power to the last breath. Although it cannot be expected that Iraq would sustain a massive military intervention by the U.S. for long, such an armed conflict would be much different from that of the 2nd Gulf War or of Afghanistan and Kosovo.

First, even if Iraq shows readiness towards fulfilling U.N. demands, none of the Arab states will support U.S. action against an Arab brother state. Apart from the U.K. and some Emirates, the U.S. would have to rely on itself and would thus experience a rude awakening in an angry environment of Arab nations. It can be easily foreseen that a "Western" intervention of Arab soil would also provoke new outbreaks of fundamentalist terrorism against Western targets.

Secondly, there is no "northern alliance" to rely on to march into Baghdad. The Shiites in the south, who used the protecting cover of the Gulf Alliance in 1991 for an uprising, do not provide a stable political prospect for Iraq. Even the U.S. does not show enthusiasm in this respect, especially because of fundamentalist tendencies among the Shiites and in the context of the critical relationship between Iraq and Iran. The Kurdish minority in the north is also not a plausible "ally". Turkey may make use of a military intervention by the U.S. from the north to search for a kind of a "final solution" to the "Kurd question". Ankara has already stated that the oil fields in the north of Iraq are actually located on "Turkish Territory". Any political option for Iraq would have to take the Sunnites into consideration, otherwise a dismemberment of Iraq with unpredictable consequences for the stability of the whole region could occur. And, finally, a regime that is threatened with death may feel provoked to expand the price for an aggressor to the maximum because it has nothing to lose.

Overthrowing a government with military means, apart from the political, and legal aspects mentioned above, and the immoral aspect of killing innocent people in the course of a war, also has to be questioned from a military point of view – whether the possible risks of such action surpass the risks of refraining from military force. This also holds true when the only solution in sight is the elimination of allegedly existing nuclear weapons. Namely, if Saddam Hussein has nuclear weapons, these weapons can not be dismantled by using force, without running the risk of having the Iraqi regime use these weapons as a last resort when threatened with death. In conclusion, the two principle goals of any concerted military action against Iraq, of either toppling the regime or establishing a stable post-Hussein government and the destruction of allegedly existing WMD, cannot be achieved by military means without putting at risk the lives of innocent people, the integrity of the Iraqi state, regional stability in the Near and in the Middle East, the environment (see: the burning of oil fields in 1991), the international oil market, and combating international terrorism.

IV. Preliminary conclusions

Proactive military action against Iraq by individual U.N. member states would be illegal, politically unwise and highly risky in military terms. As far as war and the legal character of peace enforcement is concerned, a substantial reform of the U.N. Charter regulations regarding Article 2 is apparently overdue. At the core of such a reform should be the monopolizing forces of the U.N. The U.N. must be qualified to become capable to support its own Peace Enforcement Resolutions under Chapter VII. Such steps could include, *inter alia*, the establishment of U.N. troops at the disposal of the U.N. Secretary General upon order of the Security Council, as once proposed by the former Secretary General Boutros-Boutros Ghali (Agenda for Peace). The legitimacy of military action under the provisions of the Charter should be solely left to the precise and formal decision-making of the Security Council.

In political terms, the situation in Iraq has to be handled with approaches which include regional stabilization strategies. The given resolutions by the Security Council provide enough leverage to develop a multilateral and complex crisis resolution scenario. Regional arms control and a highly effective non-proliferation strategy should be the integral parts of it.

Finally, a strategy that relies primarily on military pressure is doomed to fail. It may even create more instability than that which already exists in the region. Military means should be used as a last resort and cannot replace the lack of political concepts. It is clear that if a credible concept existed, incentives for using military power would have probably disappeared. As long as no credible political concepts exist, any military calculus should be dropped from the operational agenda to solve the Middle East crisis.

Is Attacking Iraq a Good Idea?*

John Langan

Down the street from your house is an unpretentious bungalow. You don't often see the man of the house. He wears dark suits and dark glasses on the cloudiest and the hottest days. You sometimes notice bulges in his clothing. He rarely speaks or shows much interest in the neighbors. The women of the household are lovely but have a very quiet and subdued manner. The children occasionally play in their own backyard but never speak to the neighbors. Strangers who dress and act in the style of the man of the house occasionally visit but move on quickly. Other strangers visit but are never seen to emerge from the house. Late at night loud noises and screams are sometimes heard from within the house; at other times a loud soprano voice can be heard. Swastikas and bloodstains are glimpsed on those very rare occasions when deliverymen or curious neighbors are able to get a glimpse of the interior. Altercations with the neighbors are rare but heated and usually end in sullen silence after he issues vague but nasty threats. Trucks carrying chemicals are often parked in front of the house, and large unidentified packages are brought inside. Two Dobermans appear behind the fence; razor wire is placed on top of the fence. Shrubs on neighboring lawns have been removed in the wee hours of the morning. Smoke comes out of the chimney during a long heat wave. Strange odors emanate from the house and permeate the neighborhood. Rumors are repeated that he is often seen at a nearby reservoir. Stories circulate that the man of the house is a recently released convict with a long history of criminal activity. What are the neighbors to do?

---

* First published in: America, September 9, 2002. Reprinted with permission from America Press, Inc.

The standard answer in the American suburban setting is to contact the police who, if they think the matter should be taken seriously, will investigate the man and the house, obtaining a search warrant if they have reason to believe that a crime has been committed. The police will conduct a search using no more force than is necessary and will arrest the man if they determine that a serious crime has been committed. The man will be tried and, if convicted, imprisoned. The questions about his activities and his purposes which so agitated the neighborhood will be answered in the course of the trial, and the threat which he posed will be removed. If the man refuses the search warrant or resists arrest, there may be a difficult interval, perhaps even a siege or a shootout, until overwhelming force is brought in and order is restored. Harm to the neighbors is prevented, and the threat from a dangerous criminal is removed.

Something on these lines seems to be the treatment which President Bush plans for Saddam Hussein. The point of concern for the advocates of a pre-emptive war against Baghdad is the need to eliminate a dangerous tyrant who is accumulating weapons of mass destruction, who supports terrorist activity, and who has a long record of aggression against his neighbors and cruelty against his subjects. This goal seems to be in accordance with our sense of justice and with the requirements of prudence in a dangerous world. It constitutes what in the tradition of just war thinking would be counted as a just cause; for it is a plan for defending our country and our ally, Israel, against the likelihood of an indiscriminate attack using weapons of mass destruction. But the plan serves as part of a utilitarian justification for using force and as one reasonable specification of the demands of "Realpolitik" in a very contentious part of the world. To borrow from our suburban analogy, we can readily agree that the street would be better off if everyone knew what was going on inside the dark house, if the weapons were removed, and if the dangerous resident were eliminated, and that the world would be better off if somebody had the clarity of purpose and the power to see that these things were done. For many people, including, one suspects, the president, that settles the matter. Removing Saddam Hussein should be seen as a service to the international community, to the states which neighbor Iraq, and to the people of Iraq themselves; it should also be seen as a legitimate move to defend the people of Israel and the United States against terrorism.

But there are numerous other issues which this little parable does not bring to our attention, and they are issues which have to be dealt with before we can conclude that a war against Iraq is reasonable and morally justifiable. The American people and their leaders are not being asked to give an up or down vote on Saddam Hussein or even to endorse one proposal for eliminating him. There are other matters about which we have to think as carefully as we can.

First, in the neighborhood story there is a well defined authority to which appeal can be made to investigate the situation and to enforce the law once it is determined that the man in the house has violated it. In the case of Iraq the United Nations can perform these tasks to some extent; but since 1999 Saddam Hussein has refused to cooperate with the U.N. inspection teams. Before then he made repeated efforts to frustrate and evade their activity. It does not seem that the Bush administration wants to turn to the United Nations for working out the further stages of this problem or for authorizing the use of force against Saddam Hussein. Given the likelihood of a Chinese veto and the reservations which many observers have expressed about what we propose to do, it is very doubtful that U.N. authorization for the removal of Saddam Hussein can be obtained. In the absence of such authorization, efforts to remove him come much closer to a form of vigilante justice in which considerations of necessity and urgency outweigh the requirements of proper procedure and national sovereignty. Vigilante justice may do the right thing in some cases, but it does not have the stable support which normally results from reliance on orderly procedure and processes, from the inclusion of parties which are not directly involved in the original dispute, from efforts at rational persuasion, and from the legal enforcement of norms recognized and consistently affirmed by the relevant community. Vigilante justice or "taking the law into our own hands" always gives rise to questions about whether justice is really being done precisely because "justice" is being administered by aggrieved and angry parties and because it neglects the need to ensure that justice must be seen to be done. Persuading the international community about the rightness of what we propose to do is not merely a desirable additional feature of policy; it is an essential step in ensuring that what we do serves to establish and maintain a just international order over time.

Second, the parable ends with the removal of the villain from the neighborhood. The scenario is short and happy. There is a peaceful suburban situation which can be easily restored by the removal of the bad guy. There is nothing like this in the future of Iraq, which is a country divided by two unresolved civil wars – one with the Kurds in the north, the other with the Shi'ites in the south. It is not easy to foretell what the balance of power would be between those officials in the present regime who may change sides in a timely fashion and the exile leaders, many of whom have been out of the country for a long time and who lack well defined political bases. One of the neighbors, Turkey, is adamantly against enhanced political power for the Kurdish minority; another, Iran, has strong religious sympathies with the Shi'ite rebels in the south, but will be extremely unwilling to accept a U.S. imposed settlement of the region's problems.

Third, in the neighborhood situation there is a clear and well defined set of norms about what people may or may not do in their houses, about how they may treat their neighbors and their families, and about what weapons they may possess. It is not clear just what set of norms the United States wishes to establish for the possession of weapons of mass destruction in the Middle East. What seems to guide policy is an attitude which accepts the possession of such weapons by people with whom we feel comfortable and whom we trust (the Israelis) and by people whom we need (the Pakistanis) and which denies them to regimes which we regard as hostile (the Iraqis, the Iranians). But this stance cannot lead to a generally acceptable set of norms, and it is liable to embarrassing shifts when alliances go sour and when regimes are overturned. The withdrawal of the United States from the ABM treaty earlier this year did not produce the crises which some observers had predicted, but it is a turn away from a regime of fixed norms and broad international consensus for handling the problems created by weapons of mass destruction. The United States is in the awkward position of insisting that the problem of weapons of mass destruction is sufficiently urgent to justify overriding national sovereignty by force even while it is clearly unwilling to accept any significant limitations on its own sovereignty in this area. The negative possibilities resulting from the deployment and use of weapons of mass destruction and from the diffusion of knowledge about how to make and use these weapons are, however, so great and so lasting that a plan which offers a quick solution for one crisis is not worth adopting if it will compound future crises. Is it reasonable for the United States government to think that its seizure of Iraq's weapons of mass destruction, assuming that this can be carried out in a reliable and expeditious way, will convince all of its likely adversaries to renounce all plans for acquiring such weapons? Is it not just as likely to convince that the acquisition of such weapons must be done in a deniable and concealable fashion?

Fourth, the story about the house in the neighborhood gives us only one side of events. It is told from the standpoint of an outside neighbor looking in. The key question is: what are we (the neighbors) to do? Nothing is said about how things look to the bad man inside the house or to those other persons in the house who may be confederates or hostages. What options are open to them? What are their plans for fight or for flight? Do they think their rights are being vindicated because they are about to be liberated, or do they think that their rights are being violated because they are about to be overrun by superior force? How much resistance is there likely to be? How persistent and how inventive, how desperate and how effective will it be? Even when one side possesses overwhelming firepower, it is obviously a mistake to regard the other side as an inert body which can be acted on, bombed, and expelled or reduced to silent and humble acquiescence. More particularly, how likely are the people defending Iraq, many of whom may love their country more than they despise Saddam Hussein, to use the weapons of mass destruction which they currently have? The bad man in the house may be quite resourceful in clouding issues and delaying action and may be very effective in persuading the other people in the house of the bad intentions of the outsiders. Furthermore, by combining threats and promises, he may well succeed in inducing an irresolution or a splitting of purposes among the neighbors themselves, some of whom may decide that they prefer to live with his ominous presence than with an unknown and unpredictable new regime next door.

Fifth, the domestic story is told with the assumption that once the bad man in the house is confronted, there is a fairly clear upper limit on the amount of death and destruction which can result. The inhabitants of the house and a small proportion of the attackers are at risk in such a scenario. The situation is very unlikely to develop in such a way that many of the rescuers or attackers fall within the power of the bad man and his confederates. The surrounding community can for the most part go about its business. But when we are thinking about the invasion of a country like Iraq, we are taking about massive operations in which a great deal can go wrong and about an arena of conflict in which even a side which is vastly superior in resources and technical systems will find it difficult to maintain constant control over the entire field of battle. Costs and casualties which can be foreseen with some accuracy in the neighborhood situation are much less easy to predict in a complex regional conflict. In the case of Iraq, which has had a powerful central government and cohesive military units such as the Republican Guard, costs and casualities will almost certainly be considerably higher than in Afghanistan or in the Gulf War. Both imperfect targeting by our own forces and fierce resistance by even a fraction of Saddam's supporters will put at serious risk to the lives of those civilians who are innocent of his crimes and whom we claim to be liberating.

For all these reasons, it is fundamentally mistaken to think that changing the regime in Iraq is simply an emergency enforcement of international norms which will be no more than an episode in the pacification of the Middle East and a step in the stabilization of American hegemony. If we are not seen to be doing justice and if we are not doing justice in a way which is intelligible and plausible to the rest of the world, we will be unable to build a lasting structure of peace which will include reliable means of controlling the spread of weapons of mass destruction and of ensuring that they are not used. More specifically, in the Middle East, what we propose to do will not be plausible unless we are able to present a convincing case for the view that Saddam Hussein is actually on the verge of using weapons of mass destruction. The credibility of such a case will, for obvious psychological and political reasons, depend on whether we are trusted to be working for a just resolution of the dispute between Israel and the Palestinians. Actions which are undertaken for the laudable purpose of defending Israel but which are perceived within the Islamic world as arbitrary injustices, as bloody humiliations, and as exercises of power without accountability will ensure that over time new versions of Saddam Hussein and Osama bin Laden will arise. Even when force is necessary and justifiable, it should not be allowed to set the agenda or to define relationships in a crucial part of the world. Before we attempt the more specific judgments on whether an attack on Iraq can be a just war, we and our potential allies need to achieve some clarity about whether such an attack fits into a coherent plan for the future of the Middle East and for the control of weapons of mass destruction around the world. My own suspicion is that it will be very difficult to show that an attack on Iraq is a good thing without a heavy reliance on wishful thinking and on implausible presumptions about the readiness of the world to applaud American righteousness. Winning such a war may well leave us farther from the long term goals that really matter, the creation of a just and stable situation in the Middle East and the maintenance of a world order which makes the use of weapons of mass destruction increasingly unlikely. The simplicities of vigilante justice can offer intense but short-lived satisfaction; they are not an appropriate way to achieve lasting goals which are essential to the security of Americans, their allies, and the world at large.

Is There a Just Cause for War Against Iraq?*

John Langan

The critics of just war theory often raise the question of whether the proponents of the theory have ever been able to provide clear guidance beforehand on whether or not a proposed war is or is not just. Conscientious officials and military personnel who are anxious that their actions meet the test of justice ask very similar questions of those of us who discuss just war theory as part of our academic work. It is, of course, easier to demand answers than to arrive at them, since wars are necessarily controversial and are fought under conditions of ignorance and uncertainty. Hindsight is genuinely useful when it enables us to clarify how our understanding of a complex conflict actually developed and how we came to form a considered moral judgment on a shifting and often puzzling reality. It is not reasonable to expect theories, however rooted in military history they may be, to dissolve the fog of combat; but it is reasonable to ask those who expound them to alert us to some of the problems that are likely to arise as we move from public debate toward the use of force and as we look at possibilities that may well become very morally troubling.

In contrast to the public discussion that preceded the Gulf War of 1991, there has not been much use of the language of "just war" in the public debate or in the administration's arguments for its position that there must be an immediate regime change in Iraq. A recent exception to this is the letter which Bishop Wilton Gregory sent to President Bush on September 13. In this article, I will not be commenting on Bishop Gregory's letter which he sent in his capacity as president of the U.S. Conference of Catholic Bishops and which has an official authority and a political weight which a scholarly

* First published in Georgetown Journal of International Affairs, Vol. 4.1 (Winter/Spring 2003).

comment cannot have. My own interest is in exploring certain questions which arise as we apply just war criteria to the current situation. In doing this, I am, of course, looking into a future about which our knowledge is quite limited. The precise way in which the war would be conducted has been the topic of vigorous speculation and of surprising leaks; but no one, certainly not the planners, can speak with certainty about how the course of a war will actually go. Many of the details of the present situation are unknown even to specialists on Iraqi affairs and U.S. military planning.

The first and most fundamental requirement that any proposed conflict must meet is that there be a just cause for which the war is to be fought. In the absence of a just cause, there can be no just war; and so this will always be the most fundamental requirement. It is here that the administration's proposal to invade Iraq in order to bring about a regime change in Baghdad runs into its first serious difficulty. There can be no doubt that this proposal aims at morally worthy and politically important objectives. Both the removal of Saddam Hussein from power in Baghdad and the removal of weapons of mass destruction from the arsenal of Iraq are compelling and even urgent goals. They are, of course, distinct goals; and two of the questions which U.S. policy may have to answer is whether one of them is more important to us than the other and whether we would be content with a final situation in Iraq in which only one of these objectives was actually attained. What will it be right for us to do if Saddam offers to give up the weapons of mass destruction but demands to remain in office? What will it be right for us to do if a military clique overthrows Saddam but refuses to accept what it regards as infringements on Iraq's sovereignty? In the actual situation, of course, the two objectives are bound up together; but both the ongoing debate and future policy decisions will be shaped by the priority we give to one or the other. If, for instance, we give our priority to the removal of weapons of mass destruction, as we have seemed to do in our efforts to get a positive resolution from the U.N. Security Council, we put ourselves under serious pressure to come up with consistent policies for handling other countries in the region which have or are about to have weapons of mass destruction. The need for such policies has been underlined by the disclosure that North Korea already has a small nuclear arsenal. In particular, if Saddam Hussein makes apparently serious offers to allow inspections, will we accept his remaining in power? But these questions should not cause us to deny the real benefits of removing both Saddam Hussein and his weapons from the scene.

For just war theory, however, the question about any proposed use of force is not whether it leaves the world better off in some respect, but whether there is indeed a specific just cause for a particular country or group of countries to use force against an aggressor or against a country which is doing great harm or is about to do so. The harm can affect us, or it can be directed to one or other of our allies, or it can be turned against a neighboring state. In all these sorts of cases, whether they be self-defense or the honoring of the terms of a just alliance or action in the cause of collective security, there is a just cause for war which is recognized by the moral tradition and by international law and which can usually be communicated and understood across political and cultural boundaries. The nearly universal recognition of the need to expel Iraqi aggressors from Kuwait was rooted in a recognition of the clear presence of a just cause for a defensive war in that case. The harm which defensive military actions try to end is normally inflicted by the armed forces of the aggressor country; but we can also envision scenarios in which the harm is produced by guerrilla or terrorist groups or by state actions cutting off such vital supplies as water or energy. The harm must also be current; harms inflicted in the historic past would not justify violent action now to undo them, even though there may be a serious case for renegotiation of the issues and for reparations for harm previously done.

The harm done to citizens and residents of the United States by the terrorists of Al-Qaeda aided by the Taliban on September 11, 2001, provided just cause for the use of force by the United States and its allies in the war against Afghanistan. But in the absence of convincing links between that attack and the activities of Saddam Hussein's regime in Iraq, it does not constitute a just cause for an attack on Baghdad. The attack on New York and Washington was a striking demonstration of America's vulnerability to weapons of mass destruction, and it probably created a broad willingness in the general public to use force against those who would bring harm to our shores or who would threaten our allies. But this is a psychological connection, not a moral argument founded on rational analysis. Some of the public statements of the administration appeal to this connection, but many of them respect the difference between the harm actually done by the terrorists of September 11 and the harm that has been done and may yet be done by Saddam Hussein and his minions.

If one accepts the conclusion that there has been no significant cooperation between Al-Quaeda and Saddam Hussein's regime, then the issue of just cause comes to be the question about the extent to which it is justifiable to anticipate or preempt an attack by Iraq against the United States and its allies, particularly Israel. A requirement that we must wait until an attack has actually begun seems unrealistic in a time when missiles can deliver destructive payloads within minutes and when terrorists can launch attacks from unexpected directions which reach their targets as a nearly perfect surprise. Some have claimed that the mere possession of weapons of mass destruction by a rogue state or by a terrorist group constitutes an intolerable threat to the security of the United States or its allies; in this view it is even argued that possession should be regarded as equivalent to use. It is clear that the acquisition and possession of such weapons indicates the presence of anxiety and hostility, and it is reasonable for a nation state that thinks that it is the likely target of such weapons to take measures to defend itself against this possibility. Indeed if the danger is grave and imminent, then the state may well be justified in attacking first.

The question then is whether the Iraqi threat to the United States is grave and imminent at the present time. The consensus seems to be that the Iraqi regime possesses chemical and biological weapons which could be used at any time and that it does not now possess nuclear weapons and the delivery systems which would enable it to use weapons of mass destruction in a reliable fashion against the United States. It has to be clear to all that any use by Saddam Hussein of weapons of mass destruction against his neighbors or against Israel or against the United States would lead to massive retaliation and that it would initiate a series of events which would have to include his removal from power and the destruction of his regime.

At this point the psychological and motivational differences between Saddam Hussein, who has been a secularizing leader and an opportunistic user of Islamicist slogans, and the radical militants of Al-Qaeda are very important. A reasonable intepretation of these differences is that Saddam Hussein is likely to be deterred by the prospect of the complete destruction of his regime and by the comprehensive damage which his country would suffer, whereas the prospect of death and destruction has a demonstrated positive appeal, at least to the most committed members of such organizations as Al-Qaeda who are ready to seek martyrdom. Deterrence does not come with an absolute guarantee of its effectiveness in preventing hostile or irresponsible acts; but it is so far the most reliable way which we have of working our way past intense conflicts between hostile and heavily armed powers. If deterrence is a reasonably reliable means of preventing Saddam Hussein from engaging in hostile acts against his neighbors, then it is rash to conclude that his acquisition of weapons of mass destruction will lead to his using these weapons, unless of course our own hostility to him is so manifest and so intense that he concludes that he is bound to be destroyed in any event and so he might as well take as many Americans and Israelis with him as possible. Our threats may, contrary to our stated intentions, make the use of weapons of mass destructions more rather than less attractive to him. This is a possibility which is not merely hypothetical since it is an essential part of the Bush and Blair reading of the situation that he has quantities of chemical and biological weapons available for use and since their announced agenda includes his destruction.

Some observers have pointed out that Saddam Hussein has not been deterred from rash and destructive actions in the past, for example, his attacks on Iran and Kuwait. He is, as I argued at the time of the Gulf War, a "serial aggressor", a man who cannot be relied on not to attack or to kill in the future. He is opportunistic, he will take unwise risks, he will attempt to exploit divisions and uncertainties in the ranks of his potential adversaries. For all these reasons, it will be a positive moment when he is ultimately deprived of power. But, since he does not seem to be directly self-destructive or to be driven by fanatical beliefs which are beyond effective criticism, it makes sense to deter him by drawing clear lines beyond which he must not pass and by making definite threats about what will happen if he does so. We should also remember that in the past there were certain ambiguities in the position of the United States with regard to Saddam Hussein because of our opposition to the Islamic regime in Iran and because of our failure to intervene to save the Shah's regime. These ambiguities have long since been eliminated and will not have the effect of diminishing the credibility of our threats to use force in suitable circumstances.

This point brings us to a matter which is not often discussed, namely that there is good reason to think that our concern is not so much to prevent Saddam Hussein from using his weapons of mass destruction (a move which would be foolish as well as evil), but to prevent him from ensconcing himself in a position where by virtue of his own deterrent capabilities he could restrict our own freedom of action in the Gulf area and in the Middle East more generally. No reasonable person could want to give Saddam a veto power over the actions of other powers in the Gulf region; but the history of the Cold War shows that it is possible to live with hostile, heavily armed regimes and to wait them out, precisely because they have enormous negative factors in their internal composition. Weapons of mass destruction, however, will appeal not merely to Saddam but to other regimes in the area, as an attractive means not of conquering their neighbors but of preserving themselves. Such a situation of regional deterrence would not be a positive development any more than the evolving system of deterrence on the Indian subcontinent between India and Pakistan is. But there is a serious case for regarding it as less destructive than a full scale war to destroy Saddam's regime. It does not rely on U.S. interventions in the area, which could well turn out to be more costly and more difficult to sustain than the American public is ready for. The conclusion I would draw is that weapons of mass destruction are for Saddam both attractive and virtually unusable.

Nor is it very likely that he will attempt to attack the United States with these weapons through surrogates such as Al-Quaeda or other terrorist groups. He will not want to let those weapons of mass destruction which he prizes highly pass out of his control. As long as he remains rational in the minimal sense which a system of deterrence presupposes, he will not want to run the risk of detection in any such scheme, which would require him to have a high degree of confidence in the competence and discipline of his confederates. In the aftermath of its successful operation of September 11, Al-Quaeda might seem to be qualified on these grounds; but given the conflicting viewpoints of the secular regime in Baghdad and the religious fanaticism of Al-Qaeda, it is difficult to imagine Saddam Hussein regarding them as reliable allies, allies to whom he would of necessity have to entrust the continued existence of his regime once he became their active collaborator. Such a step, which would require trust in political groups not under his complete control, seems incompatible with the manifest paranoia of his regime.

If Saddam Hussein and his regime can be deterred, then it seems that from a just war perspective this should be the preferred policy. This does not mean that deterrence is itself a satisfactory situation, only that it is better than a preventive war. Affirming a policy of containment does not mean that there could not be strong justifying reasons for fighting a war against Saddam Hussein at some later point in time, if he transgresses the limits of an internationally authorized system of inspections or if he assaults his neighbors. Such a justifiable war against Iraqi aggression (if and when it occurs) should not be fought as an unlimited war without regard for civilian casualties. Our exercise of force should not be measured by our capabilities, which are enormous, but by the military needs of the situation and by the moral requirement that we not directly target civilians and that we try to minimize civilian casualties.

The superiority of containment to preventive war does not mean that we should give up our efforts to impose on Saddam Hussein a system of inspections and on Iraq a future without weapons of mass destruction. The inspections which we are demanding will have to be intrusive and coercive if they are to be effective. Both effective containment and war require the deployment of military forces and a credible threat to use them. But a policy of deterrence and containment moving toward regional disarmament should be workable and is morally superior to an invasion of Iraq with the ensuing occupation of the country. Such a policy will have its own moments of danger, since Al-Qaeda will not come to love us for sparing an Iraqi regime which they have their own reasons for despising; it will also have the danger that the tedious task of maintaining inspections against a wily and determined foe may come to seem boring to the world at large and unduly harsh to those who lose sight of the character of Saddam's regime.

Adopting a policy of containment and deterrence means that we have to make a choice with regard to our objectives in our dealing with Iraq. We give the effective priority to the elimination of weapons of mass destruction; and we postpone and subordinate the objective of effecting regime change. It is clear from the recent unanimous vote in the United Nations Security Council that the disarmament of Iraq is an objective capable of producing a broad international consensus, whereas the removal of Saddam Hussein is not. The two objectives, however, are not neatly separable. As long as Saddam Hussein is in control of Iraq, there remains a continuing possibility that he will once again go back to attempting to build an arsenal including weapons of mass destruction. Some members of the Bush administration have made it clear that they think there are strong reasons for insisting on both objectives. If we were to imagine a scenario in which Saddam Hussein is overthrown from within, we would still have the task of persuading the successor regime to accept serious limitations on its sovereignty with regard to the possession and use of weapons of mass destruction. Whether it happens in one or in two stages, it seems that the Iraq of the future will be without weapons of mass destruction and without Saddam Hussein.

If something like this line of argument is right, then it seems that we lack a just cause for attacking Saddam Hussein and that the use of force is not a last resort in dealing with Iraq. But we encounter a certain paradox, which should make us aware of the limitations of just war thinking and sceptical about some of the dichotomies which people are only too ready to impose on what is in many ways a very fluid situation. For it seems that it is precisely by threatening the use of force, something which is not fully justified on moral grounds, that we have been able to concentrate the minds of the Iraqi government and of many other governments on the imperative need to terminate Iraq's programs for acquiring weapons of mass destruction. It is unlikely that denunciations of Saddam Hussein and lamentations about weapons of mass destruction and recriminations about his efforts to avoid and then to terminate inspections, no matter how numerous and eloquent, would in themselves have moved the situation forward to even a partial resolution. This, however, is one of the widely recognized problems of deterrent systems in general, namely, that they rely on morally questionable threats. In the case of Iraq there was reason to fear that the United States government believed that it would be advantageous to carry out the threats it was making and that because of its consciousness of its own great technological and military superiority it could implement these threats without exposing itself and its interests to serious damage. The U.N. Security Council resolution of November 8 expresses a consensus that the disarmament of Iraq is preferable to war and that this is an urgent task which justifies the threat of force as an appropriate and necessary means. The first of these conclusions is now acknowledged common ground for the administration and its critics; it restrains the more militant elements within the administration. Recognition of the second conclusion is an important point gained for the administration.

On the other hand, since Al-Qaeda and kindred groups of terrorists lack a definite and stable center of command and since key members of these groups are more than willing to sacrifice their lives, then it becomes much more difficult to deter them. Preventing attacks from these sources, which are less well defined and less easily targeted is a goal which will be more difficult to attain. It requires that we have networks of communication and cooperation which are effectively present throughout the Middle East as well as reliable sources of human intelligence. Such networks and such patterns of cooperative relationships require in the long run that we be seen as reliable and fair partners, ready to use force when needed but not expecting to stabilize the region by the intermittent application of overwhelming military might.

The achievement of stability in the Middle East is a task which in turn will require that we restore a perception of ourselves as even-handed in relation to the dispute between the Israelis and the Palestinians. This is not a matter of abandoning the Israelis as our allies, which would be shameful and foolish, but of preparing the Palestinians to work with both the Israelis and ourselves on mutually beneficial terms. Such a prospect may seem implausible in the present tragic circumstances of the conflict in Israel and the West Bank, but it is necessary if Israel is to achieve genuine and reliable security and if we are to have stable democratic allies in the Islamic world. In the long run, the war on terrorism, the effort to prevent the proliferation and use of nuclear weapons, and the resolution of the most acute political conflicts (among which the Israeli-Palestinian conflict is the most directly relevant) are all bound up together. Solving these problems will surely be more feasible in a world in which the United States shows itself to be capable of both firm leadership and genuinely collaborative action. The citizens and the government of the United States need to recognize that the resolution of these grave problems in the most dangerous part of the world requires that we not adopt means which may seem attractive in a contrived emergency but which will in the long run tell against our ability to achieve the lasting goods of peace, disarmament, and justice.

The United States: Legitimate Authority for War against Iraq?*

Gerhard Beestermöller

Whether or not a war against Iraq is in the offing is something that at this point (October 2002) only God knows.[1] But the American president has manifested his firm determination to proceed to the use of military force, even without a mandate from the U.N., if Saddam Hussein does not soon implement those Security Council resolutions which he has thus far resisted.

Bush is demanding from Saddam Hussein the return of the weapons inspectors and concomitantly the disarmament of Iraq demanded for years now, a commitment not to support terrorism, an end to the oppression of his own population, the release of prisoners of war, aid in clarifying the fate of those missing, release for repatriation of the mortal remains of soldiers killed in battle, the return of stolen property, paying for war damages as well as ceasing to trade oil outside of the limits imposed by the U.N.[2]

The American president's demands boil down to comprehensive disarmament as well as the basic internal and external democratization of Iraq under the auspices of the U.N., preferably, but of the U.S., if need be. President Bush has himself made this clear in his speech before the U.N. General Assembly on September 12, 2002:

---

* The author is grateful to Langenscheidt Übersetzungsservice for the translation.

[1] Closing date for submission of this present article was October 20, 2002. For a more thorough discussion of the thesis developed here, see my monograph *Krieg gegen den Irak – Rückkehr in die Anarchie der Staatenwelt? Ein kritischer Kommentar aus der Perspektive einer Kriegsächtungsethik*, Stuttgart, 2002 (Beiträge zur Friedensethik, no. 35).

[2] Cf. "Bush hält den Vereinten Nationen die gebrochenen Versprechen Saddams vor", in *Frankfurter Allgemeine Zeitung*, September 9, 2002, p. 2.

"If all these steps are taken, it will signal a new openness and accountability in Iraq. And it could open the prospect of the United Nations helping to build a government that represents all Iraqis – a government based on respect for human rights, economic liberty, and internationally supervised elections."[3]

Can Saddam Hussein be expected to comply with American demands? Anyone asking whether Germany should participate in a war against Iraq under the leadership of the U.S. must deal with the problem of whether and under what conditions such a war can be considered justified. Basically, this problem can be split in two, thus asking whether a war against Iraq is objectively justified, whether it thus remains the only means of last resort to ward off a unique unacceptable threat and whether such a war can really be waged successfully and whether it really entails the lesser evil when compared to the evils threatened by Saddam Hussein.

But another issue is, who has the legitimate authority to wage war against Iraq. In his speech before the General Assembly, the American president sounded convinced that the United States has legitimate authority to go to war against Saddam Hussein if the United Nations grants it that mandate. Moreover, Bush stated his conviction that the U.S. would also have the legitimate right to wage war on Saddam Hussein even without a Security Council mandate. If, more specifically, the Security Council does not grant any mandate, even though Bush once again and for all the world to hear developed his irrefutable compelling reasons and although his administration worked hard with a willingness to compromise for a corresponding resolution in the Security Council, then the failure to adopt such a resolution can only be considered as a failure of the Security Council. By failing, the U.N.'s monopoly on the legitimacy of force, as one would have to understand Bush's thinking, would in such a case lapse. The U.S. may then and must in fact go to war for the sake of maintaining the U.N. order and the authority of Security Council resolutions. This emerges from the following passages:

"The conduct of the Iraqi regime is a threat to the authority of the United Nations, and a threat to peace. Iraq has answered a decade of U.N. demands with a decade of defiance. All the world now faces a test, and the United Nations a difficult and defining moment. Are Security Council resolutions to be honored and enforced, or

[3] http://www.whitehouse.gov/news/releases/2002/09/20020912-1.html.

cast aside without consequence? Will the United Nations serve the purpose of its founding, or will it be irrelevant?

(...)

My nation will work with the U.N. Security Council to meet our common challenge. If Iraq's regime defies us again, the world must move deliberately, decisively to hold Iraq to account. We will work with the U.N. Security Council for the necessary resolutions. But the purposes of the United States should not be doubted. The Security Council resolutions will be enforced – the just demands of peace and security will be met – or action will be unavoidable.

(...)

We must stand up for our security, and for the permanent rights and the hopes of mankind. By heritage and by choice, the United States of America will make that stand. And, delegates to the United Nations, you have the power to make that stand, as well."[4]

In other words, however the Security Council wishes to act Bush seems convinced that the U.S. from the American president's point of view has legitimate authority for war against Saddam Hussein. My thesis, on the contrary, is that Bush's speech before the U.N. General Assembly implicitly contains a set of requirements on legitimate authority for a war which the U.S. itself does not meet even if it were to be granted a U.N. mandate. The argumentation expounded by the American president specifically boils down to saying that the international order of the United Nations can only provide a state with the legitimacy to use force if that state has shown itself to be a promoter of peace and justice in all of its policies. But this is precisely what the Americans are not doing; they are acting in relation to Saddam Hussein as representative of an international order from whose restrictions they endeavor to free themselves as much as possible. In this way, a war against Saddam Hussein, even if it be called for objectively, cannot under U.S. leadership be considered legitimate in the sense that it would be enforcing the rule of law. Indeed, one would even have to say that the U.S.'s behavior aims at jeopardizing the process of creating an international legal order.

Does this then mean that one must let Saddam Hussein do what he wants if he can only be stopped by war because there is no authority willing, in a position and ethically legitimized to do so? It is only against the

[4] http://www.whitehouse.gov/news/releases/2002/09/20020912 – 1.html.

background of questions of this type that the horizon of those issues of peace ethics with which we will have to deal increasingly in the future open up for me. May states, or may the U.N., collaborate in any way with a war which jeopardizes the advancing subordination of international relations to law and thus is to that extent unjust but still constitutes the lesser evil compare with to what would have to be tolerated if the war did not occur? What is the basis, and where are the limits, of *cooperatio in malo*? These issues can only be broached here.

In order to provide a basis for my theses, I intend to proceed in three steps. In a first step, I will show that the U.N. order is itself not completely powerless in relation to the world's only remaining superpower. By appearing before the General Assembly, Bush is acknowledging this power. But at the same time he makes it clear that this power is only legitimate to the extent decisions by the Security Council can be legitimized by standards that can be universalized and that the U.N. order loses its legitimacy the moment when the Security Council's decisions can no longer be legitimized (I.).

But when Bush challenges the binding nature of the Security Council's decisions according to legitimacy claims he must allow the very same questions to be posed to precisely that challenge. Bush is thus raising the claim that the U.S. is legitimized to put the Security Council on trial before world public opinion as the organ of which and on behalf of which the monopoly on legitimizing force is to be taken away from the Security Council, to be usurped by itself and the U.S. is to be given a mandate. This claim can only make sense, if at all, if the U.S. in all its policies is a demonstrably reliable guarantor and motor of law in international relations, as will be discussed in a second step (II.).

Precisely the present administration, as will be shown in a third step, does not meet this set of requirements. It pursues the policy of getting out of existing legal obligations and never allowing any power to come about which would be equal or superior to the United States. The U.S. must thus, if measured according to its own implicitly expressed criteria, even be denied the authority to wage war on Iraq if it should get a U.N. mandate to do so. But this means that the president is arrogating the authority of law to himself where he in reality wishes to bring about an order which cannot be considered a legal order. Does it follow from this that the Security Council may in no case grant such a mandate in order not to squander its own authority as a guarantor of peace and the rule of law (III.)?

I. The Security Council as a "legal enforcement authority"

It is part of the essence of law that it is administered in a non-partisan manner by an authority to which all at whom that law is directed are subject and to which it can enforce its will in relation to everyone. Not even international law can totally do without enforceability in a non-partisan manner.[5] Seen from this perspective, the U.N. order is still very far from being an international legal order in the complete meaning of the word. This is true for at least two reasons.

The first reason consists of the fact that the U.N. possesses no organ which can be perceived of as being an organ to safeguard law in the narrower sense of the term, since the U.N.'s main organ for the preservation of peace is the Security Council. It possesses a monopoly on legitimizing force to preserve peace. But the Security Council consists of states which have their own interests and which are thus in a permanent conflict, more specifically between their own interests and their obligations to the common good. In addition to that, the Security Council's discretionary license is very broad. The danger cannot be dismissed out of hand and reality demonstrates it as well that the Security Council's members repeatedly abuse their authority to pursue their own interests instead of serving the common good, which would normally be their task.

The second reason for calling the U.N. order's legal character into question is the U.N.'s inadequate enforcement capacity. Up until now it has not had its own armed forces. It must rely on states being prepared to carry out a military mandate granted by the Security Council. Thus even if the Security were to carry out its assignment, it is by no means ensured that its resolutions are also carried out. In addition to that, the Security Council has in principle no possibility of adopting a resolution against one of its permanent members, since every resolution requires their consent. For that

[5] In principle, a distinction would have to be made here between the enforcement of law inside a state and at the level of international law. What is decisive is that even international law cannot completely dispense with enforceability. Cf. hereon Gerhard Beestermöller, *Die Völkerbundsidee. Grund und Grenzen der Kriegsächtung durch Staatensolidarität*, Stuttgart, 1997 (Theologie und Frieden, no. 10), pp. 9–83.

reason, the U.N. order can never be enforced against those five states in the strict physical sense of the term.

But this does not mean that the U.N. order could completely and in every respect only be binding on states in the sense of a moral auto-obligation. It moreover possesses a certain "capacity to enforce". The latter becomes obvious if one remembers Hannah Arendt's insight that violence must rely on power. A group of persons who have in mutual understanding agreed on joint action has power.[6] Violence initially only constitutes a means. Means of violence are tools of human strength. Means of violence lose content where there is no power. Means of violence can only come to be deployed if human beings do not reject the equivalent orders and instead carry them out.[7] Violence thus presupposed power. The latter rests on consent.

The truth of this insight by that great German-Jewish philosopher has been confirmed precisely by the American president's address before the U.N. General Assembly. The American president demonstrates before world public opinion that he is firmly determined to provide implementation to the Security Council's resolutions, including even if he receives no mandate to do so. Crucial for the purposes of argumentation pursued here is that Bush did not, for instance, take recourse to a state's right to preventive self-defense if the Security Council does not ward off a threat to the peace with the appropriate means, as one could perhaps have expected. Rather, the American president claims the right and obligation on the part of the U.S., if required, to enforce for the Security Council its own resolutions by subsidiarity so as to ensure its authority and preserve peace and security for humankind. This emerges from the passage already recited in his speech before the General Assembly.

"We must stand up for our security, and for the permanent rights and the hopes of mankind. By heritage and by choice, the United States of America will make that stand. And, delegates to the United Nations, you have the power to make that stand, as well."[8]

---

[6] "*Power* corresponds to the human capacity not only to act or to do something, but to join together with others and to act with them in mutual agreement." Hannah Arendt, *On Violence,* quoted here from the German translation, 13th ed., Munich, 1998, p. 48.

[7] Arendt, *op. cit.*, p. 50.

[8] http://www.whitehouse.gov/news/releases/2002/09/20020912 – 1.html.

Why does the American president appeal to the U.N. at all? Why does he seek its mandate if he wishes to confront Saddam Hussein with the alternative of either meeting the conditions or being expelled from office without such a mandate? Why does he go before the Assembly with its sessions broadcast worldwide instead of addressing the members of the Security Council behind closed doors? Why does he advance the claim to defend the U.N.'s authority even without a U.N. mandate instead of claiming the right to be allowed to defend his country's security if the Security Council fails in its mission to preserve world peace?

The answer is clear: Bush must justify himself to world public opinion in order to win over the United Nations as representative of the family of nations for a war or at least smother opposition. It need hardly concern us here whether and in what way he is really interested in world public opinion, public opinion in his own country or in potential alliance partners and beyond. Nor is it necessary to go into the question to what extent endeavoring to obtain consent by the Security Council as the representative of world public opinion is merely instrumental for him with a view to any purpose whatsoever. Just as little as the issue need be pursued here on what cluster of factors the consent of other states in the U.N. depends. Crucial for my thinking is that appealing to peace and the rule of law with potentially universal arguments is not completely without significance here.

In other words, Bush can only successfully use force if he possesses enough power, thus if he finds enough support for a war in his electoral constituency and in other countries. An essential element in gaining this support lies in a mandate by the Security Council. With such a mandate a war against Iraq would not just be formally legalized. Since the Security Council has been entrusted with preserving world peace, a mandate for a war against Iraq at least potentially has the ability to legitimize Bush's war project comprehensively. It is not least of all because Bush needs such legitimacy that the protracted negotiations on a corresponding Security Council resolution came about. The legitimizing authority is a decisive bargaining chip for the other members in relation to the sole remaining superpower.

Of course, no-one will really be able to stop Bush from waging war against Saddam Hussein if the Security Council refuses to give him a U.N. mandate. But this insight does not disprove the thesis that the Security council has some "enforcement capability". It is precisely Bush's concern that he obtain the legitimacy which the Security Council can provide him with even if the latter denies him a mandate. This occurs, he apparently believes, by expounding before the entire world good reasons for the legitimacy of his concerns and thus the illegality of refusing to give him his mandate. By proving that the Security Council is letting itself be guided by considerations not based on the common good while he has precisely the common good in mind, the world will no longer see itself as being adequately represented by the Security Council and will have to fall in line behind the United States and its policies. With this argumentation the president is drawing our attention to an essential point: The authority of the Security council is not constituted by its "enforcement capability" but on the contrary it is the Council's authority that explains its "enforcement capability".

But then Bush goes a step further and not only puts the Security Council on trial before world public opinion and, if need be, will deprive it of its mandate on behalf of the family of nations. Beyond that he sees himself, if the Security Council should fail, as it were, authorized by the nations of the world by subsidiarity to be allowed or even obliged to preserve peace and justice. The United States is thus as the representative of the human race challenging the Security Council as the latter's organ of representation if it fails in its duties and, at least according to Bush, will in that way be saving the authority of that task by administering it in fidelity to its vocation.

When the president addresses world public opinion in this way in regard to the Security Council, he is raising the claim to be legitimized to do so. Would every state be entitled to behave in this way, if we are to take Bush seriously? Obviously not! What are the criteria inherent in his argumentation which a state must meet to be able to challenge the Security Council in a way that Bush feels justified in doing on behalf of the United States?

II. Criteria of legitimate authority to wage war according to the American president's speech

The American president with his speech to the U.N. General Assembly is, so to speak, putting the Security Council on trial before world public opinion. By doing so, he is making it clear that members of the Security Council in exercising their right to vote have an obligation to peace and justice. Only by complying with this obligation do their resolutions have legitimizing effect and do they have any "enforcement capability" over even the world's most powerful state. One can only cordially agree with such implications in Bush's speech.

But at the same time the American president, and this is of crucial importance, is also putting the United States on trial before world public opinion, more specifically by advancing in two respects the claim that the United States intervenes as a guaranteeing power for justice and peace in international relations: as a veto-enfranchised permanent member of the Security Council and as a state which challenges the Council to justify itself before its own forum and which believes that it possesses the authority to decide whether the refusal to grant a mandate can be considered proper discharge of its mandate, or as a failure to discharge it properly in order to be able to derive from that latter possibility a right to grant itself a mandate for war on behalf of world public opinion.

In view of all that Bush has demanded and proclaimed in his speech, a state must be extensively legitimized, and precisely according to the same standards which he uses to challenge the Security Council in front of world public opinion. Only if America meets such standards is Bush's request for a mandate or his proclaiming that the denial of such a mandate would, so to speak, be interpreted as a subsidiary mandate, internally consistent. What, therefore, are the precise conditions which Bush implicitly claims are met by the United States?

The Security Council is made up of states with their own interests which are therefore, as already stated, in a permanent conflict of interests, more specifically between their own interests and their obligations to the common good. There is therefore a possibility that the monopoly on legitimizing force, although it is there to protect justice and peace is being functionally turned into an instrument of partisan interests. This can only be prevented despite the power of world public opinion by having powerful states prohibit it themselves where they are not actually prevented from doing so by other states. In other words, the U.N. can only function as an instrument for the preservation of peace and justice if states themselves assume the obligation to pursue their interests in the context of law and to act in the interests of peace even when their own interests are not directly affected.

The self-commitment of states to peace and justice must additionally occur in a way that makes it possible for other states to follow it with proper justification. Otherwise states against whom a U.N. mandated war is directed could never be sure that it does not involve instrumentalizing the monopoly on violence. They would then be legitimized themselves to resist, even on condition that they were the source of an unacceptable threat to the peace. In addition, states that are spectators to these events would have to fear that they would in the future themselves be victimized by the U.N.'s instrumentalization. They will do everything to prevent this. This will be especially reflected in their actions to arm themselves and in the transparency with which they pursue the corresponding strategy. They might even make efforts in turn to turn the U.N. into an instrument of their own parochial interests or, if this is not possible, enforce their interests by force in violation of the U.N. order. Under such conditions the world would have return to its traditional anarchy. The U.N. can thus only fulfill its function of maintaining world peace if states themselves assume the obligation of only pursuing their own interests in the framework of law and put themselves under an obligation to act for peace and justice regardless of whether or not their own interests are directly affected.

This line of reasoning must be taken a step further. Linking the application of law to morality stands in considerable tension to the thought of rule of law since the rule of law is supposed to achieve security regardless of the good will of human beings. To that extent, this linkage can only be accepted from the perspective of setting up an international legal order as a transition phase. It follows from this that the ethos of bringing law to rein in international relations also implies more than just working for their further legal development. There must also be cooperation in the setting-up of independent legal institutions having authority and enforcement capability over every state.

The American president's announcement to restore, where needed, the U.N. order by acting in a subsidiary way for the Security council therefore only makes sense if the U.S. acts in U.N. bodies in accordance with U.N. ethics which would also include its commitment for an international legal order really deserving of that name. Otherwise it would be claiming to accomplish in a subsidiary manner outside of the U.N. order precisely what it is unwilling to do as its primary responsibility within the U.N. order. The American president is thus pleading before the General Assembly for a mandate for a state that is basically willing to limit the pursuit of its interests in the framework of law, to fulfill its obligations literally within the U.N. and to work for the creation of institutions to exercise law.

In particular, the President of the United States is advancing for his country the claim to be the advocate and promoter of justice in international relations by announcing itself to be authorized to assess whether the Security Council has forfeited its monopoly of legitimizing force and that therefore the United States can mandate itself on behalf of world public opinion in a subsidiary manner to wage war with such authority as originally rested with the Security Council.

In principle the possibility cannot be dismissed out of hand that the Security Council is letting itself be guided by partisan considerations and is failing to grant a mandate which is objectively needed. Does this legitimize a state not wishing to be prevented from fulfilling its duties due to wrongful acts of others to proceed to war? If this question can at all be answered in the affirmative, on the basis of which maxims must a state have itself defined as being fit to act as guarantor and promoter of justice? Is it legitimized according to its own indignant assessment to defend justice and peace in one place and not another, or must it in this decision be guided by nonpartisan standards which can be universalized when and where it opts in a subsidiary manner for the U.N. and goes to war?

Must not the same apply here as applies to self-defense inside the state? From the right to defend itself if state authority is not present on site, no obligation can be derived to defend others against attacks by risking lives. Why should this consideration not also be applicable at the international level so that a state may defend its own basic rights as well as that of its allies by war if the U.N. Security Council fails without having to expose its soldiers to mortal danger to defend the rights of other states?

The objection must be raised to such argumentation that domestic self-defense takes place on behalf of the state which for whatever reasons has no regular representative on the spot. In such a case, the situation appoints the private citizen as an official if he wishes to be appointed in this way. For that reason, a private citizen may in self-defense likewise kill a large number of criminals since he is not just defending his own life but the status of law as well. The self-defense does not therefore constitute any legitimate use of force outside of the order of a state of law but is to be seen as its execution.

By contrast, waging a war without a U.N. mandate that is not to be considered self-defense constitutes a violation of the international order of the U.N. Anyone waging such a war is precisely not acting with the authority of the Security Council but in violation of it. If states in such cases were to be allowed to decide on the basis of partisan considerations when they wished to act against threats to the peace and when not to do so then this would precisely be, not an act of subsidiary enforcement of law but the return of the feudal system in international relations. In that way the development of law in international relations would surely not be served.

The waging of a war not covered by a mandate from the Security Council may thus only be justified if the state in question is not being guided by partisan considerations but by standards capable of being universalized. Only a state willing to act on behalf of peace and justice in all comparable cases can possess legitimate authority to wage such a war. These maxims have been acknowledged by states themselves in the preamble to the U.N. Charter where they demonstrate their resolve "to ensure by the acceptance of principles and the institution of methods that armed force shall not be used, save in the common interest".

Indeed, one would even have to go beyond the requirement of being non-partisan. Only a state acting with the intention of promoting peace and justice can be considered as having legitimate authority to wage war without a U.N. mandate.[9] Before justifying this thesis it might best be explained briefly. Thus the case can very well occur that after non-partisan consideration action is only to be taken against a particular state because the latter has shown itself to be an exceptionally dangerous disturber of the peace and the U.N. Security Council does not grant the mandate called for in the circumstances. Let us further assume that the state wishing to act against the disturber of the peace outside of the framework of the U.N. order wants in this way to secure access to certain raw materials. Is such a use of force allowed because it is objectively called for or is it prohibited because the state waging war is pursuing illegitimate interests?

If, according to the requirement of right intention, the state taking recourse to force would also do so if there were no legitimate grounds for doing so, in other words the state being attacked not being a disturber of

[9] On the interpretation presented here of the classical criterion of proper intention for a just war, see Gerhard Beestermöller, *op. cit.* (2002), pp. 49–53.

the peace, then the attacking state would be committing an unjustified aggression. But if the attacking state were to forego the use of force if peace and justice were not endangered, then it would be legitimized in the case described. Thus if the primary intention (finis operis) constituting the intention is the defense of peace and justice, then the action is not deprived of its justification due to intentions of the state going beyond this (finis operantis).

Let us now proceed to substantiate this thesis. It does not blur the distinction between law and morality. Domestic penal law makes reference to the significance of the intention by demanding for killing in self-defense not only objective but even subjective elements of justification.[10] Killing in self-defense and murder can be completely identical in the external course of action. The subjective elements of justification only exist if the objective elements of justification be absent. Only then does the act of "self-defense" obtain. In this context it is of no import in law whether a person has resisted in self-defense because he wished to satisfy his hatred in such a way. This is a question of morality. The subjective elements of justification crucial in law surely no longer obtain if one human being kills another and thus prevents the latter from killing a third person where the first one was in no way in a position to know what he was preventing. In that case, there is surely no case of assistance to another in need.

Let us now transpose these considerations to the issue of legitimate authority for war while violating the U.N. order. Even if an unacceptable, exceptional threat to the peace emanates from a state, a war against it cannot be legitimate if it is conducted by a state which is little concerned about justice and would just as soon eliminate a competitor who was not threatening the peace. Such a war would in itself constitute a breach of law and can certainly not promote the process of forging an international legal order. Therefore legitimate authority only attaches to a state which would not conduct such a war if the state against which the war is directed were to act in conformity with law. Or expressed in another way: Under certain circumstances, a state could be justified in confronting a serious threat to peace with the means of war if it would also do the same thing if that threat were to emanate from a friendly state or even from a client state and if it

[10] Cf. Theodor Lenckner, § 32 *Notwehr*, in Schönke & Schröder, *Strafgesetzbuch. Kommentar*, 24th rev. ed., Munich, 1991, pp. 548 f.

would do so even to the detriment of its own interests. Whether or not a state acts with the proper intention in a specific case can be seen by how it proceeds in other comparable cases.

For a state to have the right authority for war outside of the U.N. order it by no means suffices that it acts with the proper intention. Such a war can only serve the cause of justice if other states have good reasons to trust that the right intention lies with the side of the state waging the war. Otherwise they might have to fear becoming themselves the victims of a military enforcement of a state's own interests not tempered by justice if they would not indeed see themselves as entitled to use war to pursue their own goals.

The trust placed in the legal anchorage of a state in a specific case can only be based on that state's comprehensive pattern of behavior. Only when a state is ready to pursue its interests exclusively within the framework of legal limitations, acts according to non-partisan standards for peace and justice and additionally cooperates in subordinating international relations to increasing legal development is it able to have the legitimate authority to start a war, if need be without a U.N. mandate. The requirement of being non-partisan and the one on proper intention thus interlock with each other.

Summing up: The American president with his speech before the U.N. General Assembly is dragging the Security council before the court of world public opinion and challenging it to give the U.S. a mandate for war against Saddam Hussein and announcing that it will consider itself authorized to go to war even if such a mandate should be refused. By basing this on invoking the Security Council's legitimacy as a guarantor of peace and justice, he is implicitly raising for his own country the claim to be guarantor of peace and justice. Only as a country meeting those conditions can the U.S. legitimately challenge the Security Council before world public opinion in order to ask for a military mandate or, upon the refusal of what the president feels is an objectively justified mandate in turn envisage a mandate and declare itself to be the human race's organ for making this judgment. If the United States does not meet this condition, then the American president's argumentation actually does not achieve its goal but rather raises considerable doubts on whether the Security Council may grant the US a mandate. Does the Bush administration really demonstrate that it is abiding in its words and deeds with the maxim of pursuing American interests only in the framework of justice, defending peace and justice in a non-partisan manner and promoting the further legal development of international relations?

III. The United States as a legitimate authority of war against Iraq?

It is an open question whether the Security council will draft a resolution which at least meets the minimal requirements set by the Americans. But the issue is whether or not the Security Council may grant the U.S. a mandate if one applies to the U.S. the criteria which Bush assumes that it fulfils. Expressed in another way: Would a Security Council mandate really legitimize a war against Iraq under the leadership of the U.S. if one were to join Bush in going beyond the legality of the Security Council's decisions and scrutinizing their legitimacy as well. In order to proceed further in this matter, the question must first of all be raised whether or not the Americans meet the demands which they make on themselves. As will be shown, they do not do so. Three questions join up with this insight: In what way does the U.S. president undermine the process of developing the law of international relations by presenting himself as a guarantor of rule of law without actually being one? In what way would the legal process be additionally jeopardized if the Security Council would still give the Americans a mandate? And does it follow from this that the United Nations may not grant the United States a mandate?

Does the U.S. come across in all of its policies as faithful to law and as a guarantor and promoter of the rule of law in international relations? According to what has been shown above, this question can be broken down into three subsidiary questions: Can the U.S. lay claim to be solving the conflict between the defense of its own interests and its responsibility for the common good in favor of the latter, at least in principle? Are there good reasons for assuming that the U.S., where it exempts itself from the restrictions of the international U.N. order on the use of violence, is guided by standards capable of universal application and would act in a similar manner in all comparable cases, in other words not make the decision on war dependant upon partisan interests?

In answer to the question about the way in which the U.S. both inside and outside the U.N. pursues its interests in the framework of U.N. ethics or ethics to develop the legal aspects of international relations much could be said. I will here limit myself to two matters connected with accusations against Saddam Hussein and for cause of which action is to be taken against him. There is first of all the participation of the Americans in precisely that behavior by the Iraqi dictator which is said to support in large measure the thesis about his exceptional dangerousness: the attack on Iran in breach of international law as well as the use of chemical weapons. As could be heard in recent months in the American press, the U.S. at that time in no way opposed the lawbreaker. On the contrary, about 60 officers were detailed to the Iraqi general staff to provide it with information on Iranian armed units and their locations. When the officers learned that poison gas was to be used, they made no objections. Supposedly because Iraq was fighting for survival.[11] Certainly reports of this kind must now sow doubts that the U.S. is now concerned with defense of peace and the rule of law.

In regard to the accusation that Iraq supported the terrorist organization Al-Qaeda, the U.S. still owes us any convincing evidence on that score.[12] By contrast, it is a well-known and undisputed fact that Al-Qaeda only came about at all with strong support from the United States. At that time, the Americans were concerned with forming troops willing to engage in anything, including terrorism, against the Soviet occupation troops in Afghanistan.[13] With what authority can the U.S. now act against its erstwhile creation after the latter have turned on their former masters? Certainly not in the name of the non-partisan rule of law.

As regards now the relationship between the U.S. and the U.N., I will limit myself again to a few aspects. How can the U.S. claim to be trying to enhance its authority if it year after year failed to pay its dues and has thus brought the U.N. to the brink of incapacity to act? What is the status of U.S. voting behavior in the Security Council when the matter is one of adopting resolutions against human rights violations and threats to the peace on the part of its client states such as Turkey or Israel?

---

[11] Cf. Alain Gresh, *Ziel Bagdad*, in *Le monde diplomatique* (German edition), September 8, 2002; also, Dieter Bednarz et al, *Das Dossier Saddam*, in *Der Spiegel*, 2002, no. 37 (September 09, 2002), pp. 104 – 117, in particular p. 112.

[12] Cf. Peter Rudolf, *"Präventivkrieg" als Ausweg? Die USA und der Irak*, SWP Studies, no. 23, Berlin, 2002, p. 10.

[13] Cf. Pankaj Mishra, *The Making of Afghanistan*, in *New York review of Books*, vol. 48 (2001), pp. 18 – 21.

How is U.S. behavior outside of the U.N. to be interpreted? Does its behavior inside the U.N. give us cause to entertain any hopes that it would be acting against Iraq without a U.N. mandate solely because the latter poses an exceptional threat to peace and the rule of law? Do we have any good reasons to assume that the Americans would not attack Saddam Hussein even if he were only insisting on his people's legitimate right to self-defense?

Any answers to questions like these are certainly always fraught with a veil of ambivalence. One must naturally also allow for the possibility that after the end of the Cold War and in view of the new challenges posed by terrorism in the United States the realization has gained ground that in the long run only the erection of the rule of law can guarantee the security and legitimate interests of the U.S. as well. That is why our gaze must henceforth be turned towards the global orientation of current American policy only in the context of which do isolated actions or behavior patterns make any sense. Is American policy on the whole geared to helping the rule of law prevail in international relations?

This is precisely where my concerns lie. The Bush administration seems precisely not be treading this path but the path to a type of American world hegemony.[14] It is conducting a policy aimed at loosening existing international law commitments and failing to participate in the advancing process of subordinating international relations to law. In this vein, the Bush administration in its report to congress on the American nuclear concept (Nuclear Posture Review) has explicitly listed Libya, Syria and Iraq as potential targets for American nuclear armed forces. This constitutes a breach of the commitment given in 1995 in connection with a resolution by the U.N. Security Council by the nuclear powers, the U.S., Russia, China, the U.K. and France. That commitment includes the promise never to use nuclear weapons against states without nuclear weapons which are parties to that treaty unless one of those states should attack it in an alliance with a nuclear power. With this reorientation of American nuclear strategy the non-proliferation regime has been deprived of one of its cornerstones.[15] Does this not legitimize states which in the framework of this regime have

[14] Cf. hereon the very alarming analysis made by Anatol Lieven, senior associate at the Carnegie Endowment for International Peace in Washington, in *London Review of Books*, vol. 24, no. 19 (October 3, 2002) quoted in http://www.lrb.co.uk, pp. 1 – 15.

[15] Cf. Robert S. McNamara / Thomas Graham jr., *Nuclear weapons for all?*, in *International Herald Tribune*, March 14, 2002, p. 8.

declared their renunciation of atomic weapons to now go out and acquire them?

In addition, George W. Bush revoked the signature which his predecessor Bill Clinton had given to the Rome Treaty establishing an International Criminal Court. The Americans oppose initiatives to develop further international law norms relating to global warming, biological weapons or women's rights. They have stated that they no longer feel themselves bound by the Vienna Convention on Law of Treaties and have thus far not ratified the 1989 Convention on Children's Rights. Finally, the new Bush Doctrine that American is entitled to wage preventive wars boils down to not having the U.S. comply in the future with restrictions in the U.N. Charter in regard to the use of force.[16]

The Americans not only make it clear that they oppose legal restrictions on their policy options and thus thwart the subordination of international relations to law altogether. In addition, everything indicates that the Americans are engaged in a rearmament policy aimed at maintaining their military superiority against the collective concentration of all other states' and institutions' military capacity. It is thus stated in the new "National Security Strategy of the United States", that "our forces will be strong enough to dissuade potential adversaries from pursuing a military buildup in hopes of surpassing or equaling the power of the United States".[17]

The policy of being so strong that it is superior to any possible adversary boils down to maintaining military superiority over the entire world, since every state, or indeed every institution and every coalition which could create it constitutes a potential enemy and to that extent all armaments as well as their potential accumulation constitute a threat. The claim to being militarily superior to everything else is diametrically opposed to the erection of an international legal order since and to the extent that law constituently includes its own enforceability. Confronted with a state that is militarily superior to all others taken together, no law can be enforced. Law's demand to be obeyed would then either dissolve into a moral imperative or into a calculation of interests. And can a state that claims for itself the right not to be subject to the authority of any enforcement, drag the Security Coun-

[16] Cf. Tony Judt, *Its Own Worst Enemy*, in *New York Review of Books*, vol. 49, no. 13, August 15, 2002, pp. 12 – 17.

[17] Quoted on the front page of the International Herald Tribune, September 21 – 22, 2002.

cil before the court of world public opinion and challenge it to restore its own authority by enforcing compliance with its own resolutions? Can the process of subordinating relations to law be promoted by such a policy? Certainly not!

The American president pleads before the U.N. for a mandate for the U.S. as a state wishing to enforce the rule of law in all of its policies and not just in certain cases or even less on the basis of a purely transient coincidence with its own interests. But it is precisely those conditions that the United States does not meet. If the Security Council can legitimately only grant a mandate to a state fulfilling implicitly the conditions Bush set then the Security council would have to deny the U.S. such a mandate. For that reason, the Americans, when measured by their own standards, can naturally not have the legitimacy to go to war against Saddam Hussein without a U.N. mandate.

In my opinion, one would have to go even further and clearly see how much Bush has damaged the process of subordinating international relations to law with his speech before the U.N. General Assembly, because that process ultimately depends on states obligating themselves and on at least minimal mutual trust in the other states taking equivalent action. That trust is impaired when a state so massively presents itself as a representative of the rule of law and peace that it even lays claim to the right to be allowed to judge the Security Council's conduct of its affairs while in reality being solely interested in establishing a new type of world hegemony.

Nonetheless, one may still wonder if there could not also be good grounds for granting or offering America a mandate in the event that Saddam Hussein does not comply with certain conditions. Why should the Security Council not grant the Americans an objectively needed mandate as long as it discharges its functions in a proper manner and acts in a non-partisan manner against breaches of the peace in a non-partisan manner. Is not the United Nations to be considered the real prime mover, the legitimate authority, whatever is to be said about the Americans?

Thomas Aquinas takes up this issue in dealing with the case of how to assess when a master out of pure compassion commissions his servant to bring alms to a needy family but where the servant seizes the opportunity to humiliate the recipients.[18] For Thomas the servant is in this case acting

---

[18] Cf. Sth. III, q. 64, a. 10, ad 3. See hereon Gerhard Beestermöller, *Thomas von Aquin*

in a reproachable manner but due to his indecent behavior the gift of the alms is in no way impaired in its moral value, since the servant is acting, so to speak, as an instrument of his master because the latter has the power to give him orders and to punish him otherwise. The United Nations, however, possesses no such power over the United States. A U.N. mandate would have to be interpreted as a mandate to set up precisely such a hegemony as Bush is striving for.

Still, if one follows my analysis, the issue still remains of whether the U.N. should grant a mandate for a war against Saddam Hussein, since it could well be that the threat to peace and the rule of law emanating from Saddam Hussein, if not stopped in time, is greater than the one emanating from the U.S.[19] One must also ask whether the prospect of a possible U.N. mandate in the event that Saddam Hussein refuses to cooperate with the resolutions of the Security Council might not deter the U.S. from starting a war abruptly and, since Saddam Hussein's giving in cannot be excluded with certainty, then perhaps a war could be prevented.

What is involved here systematically is the issue of the basis and bounds of cooperatio in malo with the aim of preventing in this way at least an even greater injustice. By what maxims should a state or an institution be guided which strives for the subordination of international relations to law but is forced to observe that it itself does not have the power to control that process and must therefore inevitably get itself involved in compromises? Which compromises are necessary and legitimate for the sake of such subordination to law and which are not? Which are forbidden because they might serve the historical advance of law but would instrumentalize individual human beings?

These are extremely difficult questions. It is without a doubt of eminent importance to clarify whether a war against Saddam Hussein is justified objectively. In the long run, however, it seems to me to be the issue of legitimate authority which is even more important. An objectively justified war becomes illegitimate if it boils down to leading the comity of nations back into its traditional anarchy.

---

und der gerechte Krieg, Friedensethik im theologischen Kontext der Summa Theologiae, Cologne, 1995 (Theologie und Frieden, no. 4), pp. 88 ff.

[19] Cf. Beestermöller, *op. cit.* (2002), pp. 15–27.

What we now need, in my view, is ethics of transition containing normative standards for states wishing to make themselves available to serve the rule of law. I propose that we should here speak of new preliminary ethnics. In doing so, I am following Kant's terminology. In his essay *Perpetual Peace* (1795) that great enlightener followed the structure of the then conventional peace treaties. Following on the so-called preliminary articles regulating the transition from a state of war to one of peace there come the actual provisions of peace. Kant is concerned in his preliminary articles not with ending a specific war but with the transition from anarchy to a peace built on the rule of law altogether. We are confronted with the same task today. What we lack, as far as I can judge, is a clear reform concept with normative standards for states willing to reform. Within that concept, conditions would have to be elaborated which the use of force would have to meet in order to have the process of subordinating international relations to the rule of law advanced, or at least not undermined.

Holy See Policy toward Iraq

Drew Christiansen

The role of the Holy See in international affairs is unique. The oldest diplomatic entity in the West, the Holy See has been engaged in diplomacy for more than a millennium. It has formally dispatched nuncios (ambassadors) since the sixteenth century. Contrary to popular opinion, the Holy See does not represent the government of the territory of the Vatican City State, but is rather the public face of the Catholic Church in international affairs. It is the only church, religious body or non-state actor to function this way. A distinctive presence in world affairs, it is no to be confused with a non-governmental organization (NGO).

While sometimes a signatory of United Nations protocols and often a negotiating party in drafting such agreements, especially at U.N. summits, the Holy See does not have a permanent representative at the U.N., but rather a non-voting permanent observer. This special status indicates that as a religious body the Holy See stands somewhat apart from the purely political interests of states. Its principal concerns today are religious, moral and humanitarian.

I. The Evolution of Vatican Diplomacy

The diplomatic activities of the Church today are far different than they were in the first half of the 19th century. Before the fall of the Papal States in 1870, the Holy See was intertwined with the Papal States, part of Metternich's Concert of Europe as a state among states. For more than half a century beginning in 1870, though popes continued to exchange diplomats with other powers, the Vatican was in the political wilderness, frozen out of international affairs by agreement of the Great Powers, especially Great Britain and the United States.

After the signing of the Lateran Treaty in 1929 with the Kingdom of Italy, the international position of the Holy See began to normalize, but only after the Second World War did it emerge an active player in world affairs. During the Cold War, especially in the pontificate of Pope Pius XII, the Holy See played a significant role in defense of Western Europe against the advance of Communism and in support of the oppressed Christians of Eastern Europe.

Following the Second Vatican Council (1962–65), however, the Church, while embracing the modern world, undertook to re-shape its diplomacy. Freed of "the ball and chain" of temporal power, the Holy See re-defined its role in international affairs as a voice of humanity and conscience.[1] In particular this meant the promotion of peace, the defense of human rights and authentic development, and support for activities and movements that foster the unity of the human family.

The Council emphasized that the Church's service to the world consisted above all in the promotion of unity, of which peace is a part, and the defense of human rights.[2] In addition, the Council mandated the establishment of what became the Pontifical Council for Justice and Peace and the network of diocesan justice and peace (and human rights) commissions.[3] It also made defense of human rights integral to the responsibilities of bishops.[4] The Helsinki Accords (1975) of which the Holy See was a signatory, especially "Basket IV" providing for review of human rights, also helped open a new era of human rights diplomacy in which the Church energetically participated.

---

[1] Msgr. Celestino Migliore, "The Nature and Methods of Vatican Diplomacy", in Origins, February 23, 2000, 539.

[2] See Vatican Council II, The Pastoral Constitution on the Church in the Modern World (*Gaudium et spes)*, nos. 41 and 42, in David J. O'Brien and Thomas A. Shannon, Renewing the Earth: Catholic Documents on Peace, Justice and Liberation, Garden City 1977.

[3] See *Gaudium et spes*, no. 90.

[4] Declaration on Bishops' Pastoral Role in the Church, no. 12, in Walter M. Abbott, S.J. (gen. ed.), The Documents of Vatican II, New York 1966.

1. Aims in Mideast Diplomacy

In shaping its policy towards Iraq, the Holy See has had three goals: (1) to sustain the native Christian population of the region, (2) to guide moral reflection on international affairs, especially in the areas of human rights and religious liberty, authentic human development, and war and peace, and (3) to preserve the conditions of peace and help resolve conflict in the region.

2. Rome and Middle Eastern Christians

The Middle East is the historic "cradle of Christianity". In the region as a whole, there are some 12 million Christians, divided among Apostolic (non-Chalcedonian), Orthodox and Catholic churches. The largest group consists of the six million Egyptian Copts. Less than four million are Catholics belonging to several oriental churches in communion with Rome, including Maronites in Lebanon, Chaldeans in Iraq, Copts in Egypt, and Melkites dispersed throughout the region, along with smaller numbers of Latin or Roman Catholics.

Recent decades have been very difficult for Christians in much of the Middle East. While there has been little direct persecution, the general instability in the region, the lack of economic opportunity and diverse social and political pressures have converged to limit and even diminish their numbers. Today, particularly as a result of Turkey's harsh repression of the Kurds in Anatolia, the historic Christian heartland, virtually no Christians remain in that country. In Lebanon, as a result of civil war and subsequent Syrian hegemony over the country, Maronites are hemorrhaging from the country founded to give them a home. In Egypt discriminatory legal practices, e.g., strict restrictions on the building and repair of churches, and episodic anti-Christian rioting make Christians a hard-pressed minority.[5]

In Palestine, while the weak Palestinian Authority has frequently protected Christians against harassment, the Israeli occupation and the IDF's repression of the al-Aqsa Intifada has created such difficult life conditions

[5] For a general survey of the conditions of life for Christians in the Levant, see William Dalrymple, From the Holy Mountain: A Journey among the Christians of the Middle East, New York 1997; Charles M. Sennott, The Body and the Blood: The Holy Land's Christians at the Turn of a New Millennium New York 2001.

that many Christians have chosen to emigrate. In Israel official and unofficial discrimination and insensitive government policies make Christians second class citizens. Emigration is an ever-present concern and the virtual disappearance of indigenous Christians from the Holy Land seems a real possibility.

3. Council of Patriarchs of the East

One of the significant steps in strengthening the Catholic churches in the Middle East came with the founding in 1991 of the Council of Catholic Patriarchs of the East. The Council meets once a year to coordinate pastoral activities, such as inter-ritual marriages and catechesis, and to survey regional issues. The Council has also published a series of thoughtful pastoral letters on Muslim-Christian co-existence, the church, and ecumenism, as well as reflections on contemporary and political problems.[6]

In May, 1999, with the blessing of the Holy See and an exceptional latitude of action, more than two hundred bishops from across the Middle East and the diaspora communities gathered in Fatqa, Lebanon, for the First Congress of Patriarchs and Bishops of the Middle East, to consider the common future of the oriental churches.[7]

Another significant action on the part of the Holy See was the Special Assembly of the Synod of Bishops for Lebanon (1995–96) followed by Pope John Paul II's visit to Lebanon. An effort to accelerate rapprochement in Lebanon after the close of the Lebanese Civil War, the Synod laid out a strategy for a post-confessional state, which emphasized national unity and Christian-Muslim reconciliation.[8]

---

[6] On Christian-Muslim Co-existence, see "Ensemble devant Dieu pour le Bien de la Personne et de la Société: La co-existence entre musulmans et chretiens dans le monde arabe" (Bkerke, Lebanon: Secretariat General, Conseil des Patriarches Catholiques d'Orient, 1994); on the Church Mystere de l'Eglise, see "Je suis la Vigne, Vous les Sacrements" (Bkerke, 1996); on ecumenism, see "Le Mouvement Oecumenique: Que Tous Soient Un" (Bkerke, 1999).

[7] For a description of the Congress by the Council of Patriarchs, see "L'Ensemble pour l'Avenir: 'Voici, Je fais toutes choses nouvelles'." (Bkerke 1999) and "'Pour qu'ils aient la vie et qu'ils l'aient en abondance': Actes du 1er Congres des Patriarches et Eveques Catholiques du Moyen-Orient", Mai 1999 (Bkerke 2000).

[8] See Pope John Paul II's Apostolic Exhoration concluding the synod, Une Esperance Nouvelle Pour Le Liban (Vatican City: Editrice Vaticana, 1996).

Although they followed on bi-lateral agreements between the principals, the Holy See Fundamental Agreement with the State of Israel (1993) and the Basic Agreement with the PLO (2000) were further efforts to contribute to peace in the region and to the stabilization of conditions for the Christian population in the Holy Land. The two agreements also set a new model for Holy See relations in the region based on the religious liberty of believers rather than the institutional freedom of the Church in accord with Vatican Council II's Declaration on Religious Liberty (*Dignitatis humanae*) and its Pastoral Constitution on the Church in the Modern World (*Gaudium et spes*).

4. The Situation of Iraqi Christians

Accurate population figures for the Middle East are often difficult to ascertain. The World Christian Encyclopedia lists the total Iraqi Christian population in 2000 at 724,662 or as much as 3.2 per cent of the total population.[9] The *Annuario Pontificio* for 2002, the official Vatican source, records the Chaldean population in Iraq at 220, 573 across 10 dioceses. As of 1995, there were also some 81,000 Assyrians.[10] In addition, the Encyclopedia reports an estimated 265,000 Charismatic or Pentecostals, and as many as 200,000 "isolated radio Christians" with smaller numbers of Syrian Orthodox, Armenian Apostolic and Latin Catholics.

In Iraq, there has been little blatant government persecution of Christians. Government proscription of the teaching of Syriac, however, has been a blow to the preservation of Assyrian identity in Iraq. The main cause of Christian emigration appears to be the harsh conditions of life imposed by the repressive regime of Saddam Hussein, a decade of war in the 1980s and early 90s, the strict U.N. sanctions applied to Iraq in the aftermath of the Gulf War, and the economic stagnation resulting from all these factors, conditions affecting the Muslim majority as well as the Christian minority.

One should not underestimate, however, the effects of forced integration of Christians, especially Assyrians, from remote villages in the north of the country into Muslim-dominated urban areas in Saddam Hussein's ef-

[9] See D. Barret, G. Kurian and T. Johnson (eds.), The World Christian Encyclopedia, 2 vols. (N.Y.: Oxford, 2001).

[10] *Annuario* (Vatican City: Editrice Vaticana, 2002). The figures on Assyrians come from The World Christian Encyclopedia).

fort to stamp out potential points of resistance to his regime. While it was possible in homogenous rural communities for Christians to maintain their faith, it became much more difficult for them to do so in Muslim-dominated and socially complex urban areas, and so forced re-location has been a spur to migration.

The progressive alienation and dispersal of the Iraqi Christian population is also complicated by the functioning of religion as a kind of ethnicity throughout the region. The U.S. State Department's International Religious Freedom Report 2002, for example, observes that the Iraqi Government "has sought to undermine the identity of minority Christian (Assyrian and Chaldean)" groups, referring to the prohibition on the teaching of Syriac, the language of Iraqi Christians and an identifier of special significance to Assyrians, i.e., members of the ancient Assyrian Church of the East. In addition, Assyrians in the north of the country seem to be subject to isolated incidents of violence from members of the dominant Kurdish population in that region.

While Chaldeans have seemed at times to have been favored by the government, Assyrians who are of a non-Arab ethnic stock have experienced some outright discrimination and harassment. In 2002 the government passed a law placing all Christian clergy and churches under the Ministry of Islamic Property.[11] As a result of both the generally harsh conditions in Iraq and specific limitations on their religious lives, Iraqi Christians, both Assyrian and Chaldean, have joined others in emigrating from their native country. Some have also expressed fears that United States military action against Iraq will result in violence by Muslims against Christians who will be perceived as allies of the West.[12]

5. The Vatican Support for the Iraqi People and Iraqi Christians

Thus, the situation of Christians throughout the region, with the possible exception for the moment of Syria, is troubled. For this reason, the Holy See has given considerable attention to contributing to various peace processes,

[11] The United States Commission on International Religious Freedom is an independent government body that advises the U.S. president. It is distinct from the Department of State's Office of International Religious Freedom.

[12] See Rainer Lang, "Iraqi Christians Fear War Will Create Tensions with Muslims", Ecumenical News Service, November 21, 2002.

particularly in Lebanon and Israel/Palestine and moderating western policies towards Iraq, to strengthening the Catholics of the region, and building ecumenical ties with oriental Christians, concluding, for example, an understanding in 1994/95 with the Assyrian Church of the East. In the last phase (2001) of his jubilee pilgrimages, Pope John Paul II made important moves for rapprochement with the Greek Orthodox and Syrian Orthodox Churches.

Understanding that single greatest problem facing Iraq during the last decade was the humanitarian crisis exacerbated by the comprehensive economic sanctions imposed by the United Nations on the country, the Holy See focused much of its effort during the 1990s on alleviating the severity of the embargo. First, it sought through intermediaries to accelerate the approval of the U.N. Sanctions Committee to increase the flow of unrestricted goods to the Iraqi people. Later, understanding that the fate of Christians was intertwined with that of the general population, it worked for attenuation and curtailment of the sanctions regime.

In addition, Caritas Internationalis, the international network of Catholic relief and development agencies, conducted programs of direct assistance through Caritas Iraq. These have included nutrition programs for malnourished infants, well-baby programs for sustaining infant health, and water treatment programs for both urban and rural areas.[13]

II. Evolving Teaching on War and Peace

One feature that distinguishes the Catholic Church is the *magisterium*, the teaching authority of pope, bishops, councils, etc. In matters of faith and morals the church's hierarchy has the authority to teach the faithful. This includes the morality of peace and war. While the just war (has never been formally defined as a body of Catholic doctrine, it has been a matter of the ordinary moral teaching of the Church and was included in the 1995 Catechism of the Catholic Church. In fact, since Vatican Council II, Church teaching on matters of war and peace have evolved quite dramatically. The Church's use of just-war thinking has grown increasingly stringent and its support for nonviolent resolution of conflict has grown in proportion.

[13] Background data provided by Caritas Internationalis, Vatican City.

While it gave conditioned acceptance to governmental use of force under the limitations of just war in the defense of innocent populations, the Vatican Council appealed for Catholics to examine war with "a whole new attitude".[14] It gave support to conscientious objection, and it praised the witness of nonviolent activists. Since that time, a broader theology of peace has begun to emerge in official Catholic teaching and Church commentary on international affairs. The Catholic theology of peace comprises: (1) the defense of human rights, (2) the promotion of authentic human development, (3) support for international law and international organizations, and (4) a strong doctrine of human solidarity.[15]

In the field of conflict, Catholic teaching holds a presumption against the use of force and promotes nonviolent resolution of conflict.[16] The proposition that Catholic just war teaching includes such a presumption has been contested, especially by those who hold a permissive view of the just war as fundamentally providing governments permission to use force in a just cause.[17] During the 1990s, moreover, Catholic authorities also stirred controversy by endorsing "humanitarian intervention" in the Balkans, Central Africa, and East Timor. Although the Holy See used the term in a wide sense to embrace a wide range of interventions, the advocacy of the use of force to stop or prevent genocidal actions and ethnic cleansing was seen as an expansion of the category of just cause. In fact, debate over the legitimacy of "humanitarian intervention" represented a clash between Christian and state-based conceptions of just war. The Catholic tradition takes defense of the innocent as the fundamental function of the use of force and regards all political authority as oriented to that end; state-based doctrines see the use of force as tied more narrowly to fundamental state interests.

With the exception of humanitarian intervention, however, the drift of the Holy See's policy since the end of the Cold War has been to employ just

[14] See "Pastoral Constitution on the Church in the Modern World", nos. 77–82.

[15] See "Catholic Peacemaking" a summary of a paper by Drew Christiansen S.J. in the United States Institute for Peace, Special Report: Catholic Contributions to International Peace (April 9, 2001).

[16] This presumption is most clearly articulated in "The Challenge of Peace: God's Promise and Our Response" (Washington, D.C.: United States Catholic Conference, 1983), the U.S. bishops' 1983 peace pastoral, no.120.

[17] See especially James Turner Johnson, Morality and Contemporary Warfare, New Haven 1997.

war in strict fashion to criticize government's use of force and to raise the bar for its application. This was notably the case in the Gulf War (1990–91).[18] While the Pope and other Church officials in the wake of September 11 have allowed for a state's right to self-defense in the response to massive acts of terrorism, the direction of papal teaching and the Holy See's policy has been to endorse activities that contribute to peace and to oppose the resort to force, at least, as justified by the Bush Administration.[19]

The Gulf War

The shift of Catholic teaching away from the just war towards non-violence was clearly evident at the time of the 1991 Gulf War. Overall the position of the Holy See on the repulsion of the Iraqi invasion of Kuwait was cautious in the extreme. At one point, Civilta Cattolica attempted to limit the grounds for defense to resistance to "an aggression actually taking place", thereby denying justification for coalition action after the occupation of Iraq. Pope John Paul II held out against the resort to arms, which he described as "an adventure with no return" that at one point in exasperation he was forced to protest to his critics, "I am not a pacifist."[20]

The repeated concern of the Holy See in the lead up to the Gulf War was encouragement of honest negotiation between the United States and Iraq. Again and again, the Holy Father himself called for dialogue as a way to prevent the outbreak of war. On a September, 1990 trip to Central Africa, he asked that "a spirit of dialogue win over confrontation".[21] By December Italy had sent naval ships to the Persian Gulf as part of the build up for war. In a meeting with Italian naval officers, the pope prayed "for a lessening of tensions in the whole Middle East and for the settlement of the disputes through dialogue".[22] In his 1990 Christmas message, he declared, "May leaders be convinced that *war is an adventure with no return*. By reasoning,

---

[18] See John Paul II for Peace in the Middle East / War in the Gulf: Gleanings from the Pages of "L'Osservatore Romano", Quaderni de "L'Osservatore Romano", Collana diretta da Mario Agnes, no. 16, Vatican City: Libreria Editrice Vaticana, 1992.

[19] See especially Pope John Paul II's 2002 World Day for Peace Message "No Peace without Justice, No Justice without Forgiveness".

[20] "An adventure with no return", 1990 Christmas Message in Peace in the Middle East, 33–37.

[21] Ibid, 14.

[22] Ibid, 29.

patience and dialogue, with respect for the inalienable rights of peoples and nations, it is possible to identify and travel the paths of understanding and peace."[23] Speaking to the Ministers of Foreign Affairs of the European communities, he said, "the principle of equity demands that peaceful means such as dialogue and negotiation prevail over the recourse to instruments of devastating and terrifying death".[24] To officers of the NATO Defense College, he said, "we cannot forget that dialogue and negotiation and non-violence are the only effective ways for solving conflicts in full respect for the demands of justice".[25]

Days before the outbreak of hostilities, Pope John Paul wrote a pair of letters to Iraqi President Saddam Hussein and U.S. President George (H. W.) Bush. To Saddam he wrote, "No international problem can be adequately and worthily solved by recourse to arms, and experience teaches all humanity that war, besides causing victims, creates situations of grave injustice which, in turn, constitute a powerful temptation to further recourse to violence." He prayed that "all the parties involved will yet succeed in discovering, in frank and fruitful dialogue, the path to avoiding such a catastrophe".[26]

To the first President Bush he wrote, "I wish now to restate my firm belief that war is not likely to bring an adequate solution to international problems and that ... the consequences that would possibly derive from war would be devastating and tragic." He added, "I hope that, through, a last-minute effort at dialogue, sovereignty may be restored to the people of Kuwait and that the international order ... may be re-established in the Gulf area and in the entire Middle East."[27]

As the war itself unfolded papal appeals continued to appeal for dialogue, but turned as well to preventing escalation of the war and relieving its sufferings. The pope also repeatedly presented his analysis that the time for war as a remedy for injustice had passed. His most significant statement, in the encyclical letter *Centesimus annus*, praised the practitioners of non-violence, who in his view had wrought the dramatic changes in eastern Europe in 1989, and again urged an end to war as a means for the settle-

[23] Ibid, 36.
[24] Ibid, 41.
[25] Ibid, 72.
[26] Ibid, 66f.
[27] Ibid, 68f.

ment of disputes.[28] The means of destruction available even to small and medium-sized countries, he argued, make it exceedingly difficult to limit the consequences of armed conflict.[29] "May people learn to fight for justice without violence", he wrote, "renouncing class struggle in their internal disputes, and war in international ones".[30]

In a single sentence, the pontiff offered a thoroughgoing critique of war. "No, never again war, which destroys the lives of innocent people, teaches how to kill, throws into upheaval even the lives of those who do the killing and leaves behind a trail of resentment and hatred, thus making it all the more difficult to find a just solution to the very problems which provoked the war."[31] He continued, "It must not be forgotten that at the root of war there are usually real and serious grievances: injustices suffered, legitimate aspirations frustrated, poverty, and the exploitation of multitudes of desperate people who see no real possibility of improving their lot by peaceful means."[32]

So, whether we look to the consequences of war or to its causes, the Holy Father put war to question. As alternatives to war as a tool for resolution of conflict, he emphasized the primacy of international law and "a concerted worldwide effort for development".[33] "Just as the time has finally come when in individual States a system of private vendetta and reprisal has given way to the rule of law, so too a similar step forward is urgently needed in the international community."[34] In the wake of the war, Civilta Cattolica, the Jesuit journal, often described as a semi-official Vatican organ, went even further publishing a lengthy editorial trial balloon, proposing that under contemporary conditions the just war was no longer a tenable set of ideas.[35] While the proposal did not receive wide acceptance, it was a significant indicator of how Vatican thinking had been moving during John Paul's pontificate.

---

[28] On the Hundredth Anniversary of Rerum Novarum: *Centesimus Annus* (Washington, D.C.: United States Catholic Conference, 1991) nos. 23 and 25.

[29] Ibid, no. 51.

[30] Ibid, no. 23.

[31] Ibid, no. 52.

[32] Ibid.

[33] Ibid.

[34] Ibid.

[35] See Civilta Cattolica, "Modern War and Christian Conscience", in Origins 21:28 (Dec. 19, 1991) 450–455.

III. Church Teaching, Holy See Policy and the Current Debate

In the current debate over war with Iraq, whether as a preventative action to forestall the use of weapons of mass destruction or as an enforcement measure backing U.N. resolutions, the Holy See has had three principal concerns: (1) to insist on the importance of international law and the jurisdiction of international organizations, and for that reason (2) to oppose unilateral U.S. action; (3) to protect the just war tradition against attenuation by justification of preventative war, and (4) to urge disarmament and non-proliferation in general, but especially for the Mideast region.

(1) International Law. During the current crisis, comments by the Holy See on the prospect of war with Iraq have echoed positions taken during the 1991 Gulf War. Speaking in his annual address to the diplomatic corps, Pope John Paul II that year laid out the Holy See's general posture on the priority and nature of international law. On the one hand, the Holy Father acknowledged the Iraqi invasion of Kuwait as "a brutal violation of international law as defined by the U.N. and by moral law". On the other hand, he warned that putting an end to the aggression by even limited force of arms would be "costly in terms of human life". He argued, moreover, that "a peace obtained by arms could only prepare the way for new acts of violence". International law, he contended, has become "truly a code of behaviour for the human family as a whole".[36]

International law, the pontiff said, is an expression of "certain universal principles" which constitute the moral "order willed by the Creator". Among these, he listed "rejection of war as a normal means of settling conflict". Expressing the wish that international law be provided with "the coercive provisions adequate to ensure their application". Even in applying the law, "justice and equity" must be the rule.[37] This means that even in a just cause, the resort to arms must be tested by the rule of proportionality, i.e., be "proportionate to the results one wished to obtain and with due consideration to the consequences of military actions".[38] He concludes, "'The needs of mankind' ... require that we proceed resolutely toward outlawing war completely and come to cultivate peace as a supreme good ..."[39]

[36] Peace in the Middle East, 57–61.

[37] Ibid, 59.

[38] Ibid, 60.

[39] Ibid.

In the current crisis, Archbishop Jean-Louis Tauran, the Holy See's Secretary for Relations with States, continues to put proposals for coercive enforcement of U.N. resolutions against Iraq in the context of international law and the authority of the United Nations Organization. In an interview with the Italian Catholic newspaper *Avvenire*, he declared: "Should the international community, inspired by international law and in particular by the Resolutions of the Security Council of the United Nations, judge the use of force opportune and proportionate, this should happen only with a decision within the framework of the United Nations, which would decide only after having thoroughly weighed its consequences on the Iraqi civilian population, as well as the repercussions it could have on the Countries in the region and on world stability, otherwise, it would be the case of imposing the law of the strongest or of the most violent."[40] Note that in addition to international law, Tauran's remarks place the authority for enforcement measures entirely within the scope of the United Nations, thereby, denying, in this instance, the argument for "self-help" or unilateral action as proposed by the United States. (This contrasts with the Holy See's appeal during the break-up of the former Yugoslavia for any power to act against genocide and ethnic cleansing when the international community failed to take action.)

In addition, the formula on proportionality puts a heavy weight of argument even on the United Nations. Above and beyond the risks to the Iraqi population ("the question of collateral damage"), it must consider repercussions in the region and the impact on global stability. This heavy weighting of the decision against war is in keeping with the Holy Father's convictions about the evils of war and especially his judgement that war only makes it more difficult to resolve the problems for which it is the attempted remedy. Underscoring the same point, Jesuit Cardinal Roberto Tucci in an interview with Vatican Radio asked whether "certain U.S. positions ... will have the consequence ... making the expansion of terrorism and the struggle against terrorism perpetual?".[41]

(2) U.S. Unilateralism and Catholic Universalism. The Vatican's preference for multilateral action under the United Nations is, in part, an attempt

[40] *Avvenire*, 9/9/02, trans. Msgr. Celestino Migliore in memorandum to Gerard Powers (USCCB), 9/10/02.

[41] Zenit, 10/3/02.

to make the recourse to force more difficult. In this sense, it is opposed to the unilateralism announced in the White House National Security Strategy document issued last summer. But it also draws on a tradition of support for the United Nations and more broadly of international or transnational political authority as necessary for a achievement of "the universal common good". Forty years ago, in his encyclical letter *Pacem in terris*, Pope John XXIII declared his support for the United Nations system "as an important step on the path toward the juridical-political organization of all the people of the world".[42] Ever since, the Holy See has given its strong support to the United Nations and to collective efforts of "the international community". In the current crisis, Archbishop Tauran, the Vatican foreign minister, and several Roman cardinals have indicated in the most direct way that the proper site of decisionmaking authority in the case of Iraq belongs with the United Nations.

The policy favoring the U.N. coincides with some of the deeper impulses of Catholicism: its universalism, its positive reading of "socialization" and trends favoring the unity of the human family, and its endorsement of solidarity among peoples.[43] While Pope John Paul's teaching has endorsed the link between spirituality and national identity, in political ethics both John Paul and contemporary Catholic social teaching are decidedly "cosmopolitan" as opposed to nationalist or state-centered. That is, they favor the rights of human beings as such as the standard of political action rather than the interests of states. This has made the Church hesitant to enlist in military causes except for humanitarian intervention, where the rights of persons are most evidently at stake.

(3) Preventative War. The Holy See's response to debate over a new war with Iraq has come from a number of senior Vatican officials in addition to Archibishop Tauran. They have also been notable for their criticism of the concept of a "preventative" or "preemptive war", that is a war undertaken to avert a terrorist attack with weapons of mass destruction. Joseph Cardinal Ratzinger, the prefect of the Congregation of the Doctrine of the Faith, in remarks given at Trieste on September 22, indicated that while the Church continued to teach the just war, its conception did not allow for preventative war. "The concept of 'preventative war', he said, does not appear in The

[42] *Pacem in terris,* in Renewing the Earth, nos. 132 – 145; citation, no. 144.

[43] See "Pastoral Constitution", nos. 25, 42.

Catechism of the Catholic Church."[44] While certain values and people must be defended, he said, the Church "offers a very precise doctrine on the limits of these possibilities".

Cardinal Camillo Ruini, president of the Italian bishops' conference and episcopal vicar of Rome, also opposed the notion of "preventative war" in a statement September 22. The cardinal also warned that the "very great network of international solidarity which came into being immediately after September 11 (2001), seems to be cracking because of growing splits, especially in that first and traditional point of strength which is the relation between the United States of America and Western Europe".[45]

IV. Conclusion

For the Holy See, struggles in the Middle East are no ordinary conflicts. They bear on the lands that were "cradle of Christianity" and threaten the survival of the Christian presence in the region where Christianity was born. The looming crisis with Iraq, in its view, presents a special moral challenge, first because the Holy See was never satisfied that the 1991 Persian Gulf War was necessary and because it believes the Iraqi people, as a result of U.N. imposed sanctions, have suffered excessively in the interim. In addition, it fears that a war against Iraq, and especially its defeat, will stimulate a clash of civilizations between Islam and the West, giving rise to even greater terrorism and international tensions. Of course, increased ill feeling on the part of Muslims can easily translate into pressure on Middle Eastern Christians, leading to further depletion by emigration of their already reduced numbers, a real blow to the Church's own catholicity.

The Church's suspicion of war against Iraq, however, are informed by a growing resistance on the part of Pope John Paul II to war as a remedy for injustice. The Holy Father and other officials have frequently questioned whether war is a suitable remedy for today's problems, and they have underscored "the presumption against the use of force" which underlays the Catholic understanding of just war. For this reason, they constantly urge dialogue and negotiation and recourse to the United Nations. They have opposed unilateral resort to force in the matter of Iraq's alleged weapons of

---

[44] Zenit, 9/22/02.

[45] Ibid.

mass destruction and excluded "preventative war" as a means to resolve the problem.

Iraq: How Severe is the Threat?

Henner Fürtig

I. Introduction

Saddam Hussein and the Baath regime he is leading in Iraq have definitely posed a tremendous threat after assuming power in 1979. During the now 23 years of his presidency, he led his country into two wars that lasted roughly nine years altogether, and spent the remaining years to overcome the disastrous results of these wars. It should not be forgotten here that the eight-year war against Iran (1980–1988) was the longest and most devastating military inter-state conflict ever fought between two developing countries and that the war against Kuwait mobilized the largest and most powerful international military coalition against a single nation since the end of the Second World War. Saddam fired missiles against neighboring states, not only against countries he was in a state of war with (as Iran in the First Gulf War), but also against neutral states such as Israel in the Second Gulf War.

In addition, in a quasi genocidal policy, he destroyed approximately 2.000 Kurdish villages and was responsible for the disappearance of about 180.000 Iraqi Kurds in 1988 (operation "anfal"-spoils)[1] as a revenge for their alleged collective siding with Iran in the First Gulf War. And he massacred thousands of Iraqi Shiites after the Second Gulf War, not only when he crushed their upheaval in March 1991, but also when he later drained their last refuge, the marshes of southern Iraq. Nevertheless, when resuming these activities, it is obvious that the threat of Saddam's Iraq was almost

[1] See Andrew Parasiliti/Antoon Sinan, "Friends in Need, Foes to Heed: The Iraqi Military in Politics", in *Middle East Policy*, 7 (2000) 4, p. 137.

entirely directed against his own population and certain neighboring countries, but not against the world as a whole.

Even after Western politics became aware that Saddam's regime had killed 5.000 inhabitants of the Kurdish town of Halabja in a 1987 overture of the *anfal* operation by using chemical weapons, it was only taken as a proof that Iraq has deployable chemical weapons of mass destruction (WMD), but not as an incentive for a more in-depth observation. It should not be forgotten that Iraq – after the beginning of the "tanker-war" in 1987 at latest – was a partner of the West and especially the U.S. in its efforts to contain the Iranian revolution. It was only after Iraq invaded Kuwait in August 1990 that the Western perception of Saddam Hussein changed: international media described him as aggressive and power-thirsty, American politicians – including the President – compared him to Adolf Hitler, and reports that Iraq was at the brink of becoming a nuclear power were taken more seriously. Altogether, the international community became alerted and began asking questions on whether Iraq might pose a threat for the whole world.

And indeed, as revealed by one of Iraq's leading nuclear scientists, Khidr Hamzah, in 2000 after fleeing the country, Baghdad had begun to develop nuclear weapons already in the early 1970s. But until 1981, the program was heavily dependent on imported technology. After the Israeli attack at the Osirak reactor in that year, the Iraqi government intensified its efforts and invested up to 25 times more resources into the program. According to Hamzah, Iraq obtained by 1990 a "crude, one-and-a-half-ton device."[2] Apart from the fact that it is always difficult to verify reports from defectors, Hamzah's revelations came ten years too late. In 1990, international observers could only guess while Baghdad was insisting on pursuing only scientific research interests with its nuclear program. It lasted until the 1995 defection to Jordan of Hussein Kamil, Saddam Hussein's brother-in-law, who was responsible for Iraq's WMD program that the Baath regime could do nothing but to confess that it had established a comprehensive nuclear weapons program long before the Second Gulf War. According to Hussein Kamil, the program was concentrated on building an implosion-type weapon while it was accompanied by a ballistic missile project that

[2] See Khidr Hamzah, *Saddam's bombmaker: the terrifying inside story of the Iraqi nuclear and biological weapons agenda*, New York: Scribner 2000.

was intended to produce the delivery system.[3] Though in 1990, the majority of experts was maintaining that Iraq was still at least one year away from being able to produce a deployable nuclear bomb.

Although the 43 days of massive bombing raids against Iraq between January and February 1991 were believed by some to have destroyed the bulk of Iraq's WMD arsenal and capabilities,[4] others were eager to use the overwhelming allied victory in the Second Gulf War to get certainty once and for all. Therefore, they made sure that the lengthy cease-fire resolution 687, adopted on 3 April 1991 by the United Nations Security Council (UNSC), required the complete destruction of all Iraqi WMD. The resolution noted that Iraq contradicted many of its own commitments, for instance the 1925 Protocol for the Prohibition of the Use in War of Asphyxiating, Poisonous, or Other Gases, and of Bacteriological Methods of Warfare, the Final Declaration of states party to the Geneva Convention of 1989 in which the Iraqi government obligated itself to the objective of eliminating chemical and biological weapons, the Convention on the Prohibition of the Development, Production, and Stockpiling of Bacteriological (Biological) and Toxin Weapons and their Destruction of 1972, and the 1968 Nuclear Non-Proliferation Treaty.

Therefore, Iraq should declare its entire stock of illegal weapons (nuclear, biological and chemical weapons, as well as ballistic missiles with a range greater than 150 kilometers) immediately, and should "unconditionally accept" its destruction within one year. To supervise this, resolution 687 authorized on-site U.N. inspections, thus providing the legal basis for the establishment of the U.N. Special Commission (UNSCOM).[5] Article 22 of the resolution specified that the sanctions against Iraq would only be lifted after UNSCOM had declared the country free of WMD.

---

[3] See "Iraq Weapons of Mass Destruction Programs" *U.S. Government White Paper, released February 13, 1998*, Washington D.C.: Department of State 1998, p. 14.

[4] See for instance Daniel Byman, "After the Storm: U.S. Policy Toward Iraq Since 1991", in *Political Science Quarterly*, 115 (2000) 4, p. 496.

[5] See Michael Rubin, "Sanctions on Iraq: A Valid Anti-American Grievance?", in *Middle East Review of International Affairs (MERIA)*, 5 (2001) 4, p. 13.

II. Results of UNSCOM activities

Saddam Hussein refused to cooperate with UNSCOM. He neither declared his WMD nor gave any support to destroy them. UNSCOM's task became extremely difficult since the regime in Baghdad prevented the inspectors from entering certain sites and blatantly lied about even the very existence of specific WMD programs. UNSCOM chief Rolf Ekeus once stated that Iraq increased its resistance to inspections "in direct contradiction to its commitments".[6] According to an unconfirmed report, Saddam Hussein had briefed Baath party leaders that the "Special Commission is a temporary measure. We will fool them and we will bribe them and the matter will be over in a few months".[7] This prediction proved wrong but instead of cooperating with UNSCOM Saddam continued to sabotage its work using five main strategies:

"– To intimidate the U.N. by making threats and refusing to cooperate.
- To wear down its adversaries by stretching out the need to maintain sanctions over many years, when the issues could have been resolved in a much shorter time period.
- To fool the U.N. by a superficial pretense to cooperation at times and by supplying misinformation.
- To undermine the coalition by offering various countries – notably China, France, and Russia – lucrative oil, arms, and other contracts to be implemented when sanctions are removed.
- To gain support from international public opinion by depriving its own citizens of their material needs ... and blame the problem on the United States. Portraying Iraq as a nation of hungry people and sick children became a cynical propaganda tool. Blaming foreigners for the regime's decisions and mismanagement could also increase domestic support for the government."[8]

Especially the fourth of his strategies, i.e. to drive a wedge between the five permanent members of the UNSC (P5) proved highly successful. As of

[6] Rolf Ekeus, "From UNSCOM to UNMOVIC: The Future of Weapons Inspections in Iraq", Policywatch No. 477, July 19, 2000, in *Peacewatch/Policywatch Anthology 2000*, Washington D.C.: Washington Institute for Near East Policy 2001, pp. 334–336.

[7] Quoted in Andrew Cockburn, Patrick Cockburn, *Out of the Ashes: The Resurrection of Saddam Hussein*, New York, London: Verso, 2000, p. 96.

[8] Michael Rubin, op. cit., p. 14.

1997, China, France and Russia refused a robust inspections regime. When UNSCOM criticized Iraqi noncompliance, these three states demanded the end of the sanctions rather than calling Iraq to fulfill its obligations. They opposed any military action against Saddam Hussein, even when this would lead to the abandonment of the inspection system. "In terms of its rhetoric, France has always been closer to Anglo-American policies and far more committed than Russia and China to Iraq's disarmament, but in practice the differences between France and its two partners in the Security Council were small ... The French formula held that the UN should not expect Iraq to comply with its resolutions to the letter. This was in marked contrast to the Anglo-American approach that Iraq had to satisfy 100 percent of UNSCOM's demands. Likewise, while the United States and Britain sought U.N. approval for military operations whenever Iraq seriously interrupted UNSCOM's military activities, Russia, France, and China always objected."[9] In this situation, the U.S. and Britain were afraid of loosing all international approval of the sanctions system, and refrained from conducting major military actions against Iraq on their own to enforce compliance to resolution 687.

And indeed, critical voices emerged who questioned the feasibility of the inspection system at all. Instead of carrying out the very difficult, if not impossible task of accrediting all WMD including all the spare parts and all the related material within the disarmament program it probably would have made more sense to pursue a meaningful policy of eliminating Iraq's *ability* to employ WMD.[10] Such a far-reaching criticism of the inspection system remained an exception outside of Iraq but inside the country Saddam Hussein felt encouraged to test his scope of action. For instance, he initially did not accept the U.N.'s oil-for-food program adopted by the UNSC in its resolution 986 of April 1995. His propaganda machine told the Iraqi people that the U.N. – by controlling Iraq's entire petroleum transactions that way – would get a mandate over the country's most valuable resource.

---

[9] Amatzia Baram, "Saddam Husayn between his power base and the International community", in *Middle East Review of International Affairs (MERIA)*, 4 (2000) 4, p. 11.

[10] See Walid Khadduri, "U.N. Sanctions on Iraq: 10 years later", in *Middle East Policy*, 7 (2000) 4, p. 157.

This would be very similar to the political mandate Britain had on Iraq as of April 1920.[11]

Although Saddam later accepted UNSCR 986, he always tried to play the card of patriotism and national pride. Arguing that they were symbols of national sovereignty, he declared, for example, in 1997 that several presidential palaces would be off limits to inspectors – despite the fact that these sites included more than 1.000 buildings and storage rooms. In a Memorandum of Understanding signed by U.N. Secretary General Kofi Annan and the Iraqi Deputy Prime Minister Tariq Aziz on 23 February 1998, the U.N. promised to respect the sovereignty and territorial integrity of Iraq as well as its legitimate concerns regarding dignity, referring especially to the presidential palaces.[12] Against this background it does not come as a surprise that the U.S. State Department had drawn a pessimistic résumé ten days earlier in a White Paper: "On the basis of the last seven years' experience, the world's experts conclude that enough production components and data remain hidden and enough expertise has been retained or developed to enable Iraq to resume development and production of WMD. They believe Iraq maintains a small force of Scud-type missiles, a small stockpile of chemical and biological munitions, and the capability to quickly resurrect biological and chemical weapons production. This conclusion is borne out by gaps and inconsistencies in Iraq's WMD declarations, Iraq's continued obstruction of UNSCOM inspections and monitoring activities, Saddam's efforts to increase the number of 'sensitive' locations exempt from inspection, and Saddam's efforts to end inspections entirely."[13]

Thus, in the spring of 1998, the whole inspections system was in a crisis. It was obvious that the pro-Iraqi faction in the UNSC would not be able to abandon the inspection system since such a decision would immediately be vetoed by the U.S. and Britain while the latter two would not be empowered to enforce 687 by military means. Saddam Hussein on his part got the impression – or at least he propagated it – that irrespective of his or Iraq's concessions to the U.N., nothing would change the United States' and Britain's insistence on the fulfillment of UNSCR 687 while he remains

[11] See Dilip Hiro, *Neighbors, Not Friends: Iraq and Iran after the Gulf War*, London: Routledge 2001, p. 287.
[12] See Daniel Byman, op. cit., p. 504.
[13] "Iraq Weapons of Mass Destruction Programs", op. cit., p. 1.

in power.[14] Therefore, he ordered on 5 August 1998 the suspension of any cooperation with the UNSCOM regarding weapons inspections and monitoring or verification activities. At the same time, Saddam indicated that he would resume cooperation only in the case of lifting sanctions on Iraqi oil exports and a re-organization of UNSCOM's structure. The core of the latter condition was the demand to move the headquarters from New York to either Vienna or Geneva. This would make UNSCOM – according to Saddam – more independent and less susceptible of being manipulated by the U.S. On 31 October 1998, the cooperation was effectively ceased.

This time, Saddam had overplayed his card: the P5 unanimously demanded the immediate and unconditional repeal of the decisions of 5 August and 31 October. Tariq Aziz' letter to Kofi Annan of 14 November in which he announced to unconditionally resume cooperation with UNSCOM was not regarded sufficient by the UNSC because Rolf Ekeus' successor as chief of UNSCOM, Richard Butler, had on 15 December 1998 presented a report to Kofi Annan in which he complained of Iraq's rejection to full cooperation since 14 November. Under these circumstances, Butler concluded, UNSCOM would "not be able to conduct the substantive disarmament work mandated to it by the Security Council."[15] On the basis of this report, the U.S. and Great Britain started a massive aerial bombardment of Iraq the next day – without formal approval by the other P5.

During the four days of "Operation Desert Fox" Washington and London destroyed many suspicious sites identified by UNSCOM during the preceding months and years. Saddam took the chance to repeat his accusations that Richard Butler had primarily reported to the U.S. and Britain and not to the UNSC, but the air raids offered him an even greater, a "golden" opportunity to expel UNSCOM and the inspectors once and for good. He went away unpunished, except the continuation of small-scale confrontations over the no-fly zones. In the coming months, Russia, China and France, while emphasizing the need to continue monitoring Iraq's military capabilities, were trying to add more non-confrontational measures into the process. After lengthy discussions with the U.S. and Britain, all the P5 became eager to end the stalemate. On 7 December 1999, the Security

[14] See also Toby Dodge, "Winning Now?", in *The World Today*, 57 (2001) 3, p. 6.

[15] See Amin Tarzi, "Contradictions in U.S. Policy on Iraq and it's Consequences", in *Middle East Review of International Affairs (MERIA)*, 4 (2000) 1, p. 2.

Council adopted resolution 1284 which retained control over the bulk of Iraq's oil revenues, but relaxed the sanctions in some important fields. In addition, it foresaw the establishment of a new inspection body, the U.N. Monitoring Verification and Inspection Commission (UNMOVIC) under the leadership of Hans Blix.

Nevertheless, Iraq rejected the resolution because it made the full and unconditional cooperation with UNMOVIC a pre-condition for the suspension of the embargo. Hans Blix could propose the suspension of the embargo not earlier than after one year of cooperation. Iraq was also not ready to accept "suspension" since it meant that the embargo could be easily re-imposed (initially, the suspension was to last for six months only). Although Iraq rejected the resolution, no military measures were taken against the country.[16] Saddam understood the whole sanctions system as having collapsed. Thus, since the end of 1998, there were no inspections of the Iraqi WMD program within the country itself.

But it has to be stated that the activities of UNSCOM between 1991 and 1998 were not at all in vain. The U.N. inspectors unearthed – despite an extraordinary Iraqi effort of concealment – many secrets of Baghdad's nuclear, chemical, and biological weapons program. According to its report of October 1998, i.e. prior to "Operation Desert Fox", UNSCOM disposed 817 out of 819 missiles (with the fate of the other two unknown) with a range of more than 150 kilometers, was monitoring 183 sites regularly, and 52 others sporadically. The report also noted that "the disarmament phase of the Security Councils' requirements is possibly near its end in the missile and chemical weapons area but not in the biological weapons area."[17] Under UNSCOM control, tens of thousands of chemical munitions were destroyed as well as hundreds of tons of chemical weapons agent, a production facility for biological weapons, and nuclear weapons production facilities.[18] The lengthy discourse on the history of UNSCOM activities between 1991 and 1998 should therefore not depreciate the results of the U.N. body's work but explain that there might be still more gray areas in

[16] See Amatzia Baram, op. cit., p. 13.

[17] Quoted in Alain Gresh, "Iraq beyond Sanctions", in Hanelt, Christian-Peter/Neugart, Felix/Peitz, Matthias (eds.), *Future Perspectives for European – Gulf Relations*. München, Gütersloh: Bertelsmann Stiftung 2000, p. 69.

[18] See Daniel Byman, op. cit., p. 505.

Iraq's WMD program than the biological weapons area the UNSCOM is conceding itself.

III. Iraq's WMD program at present

Experiences since 1998 have provided sufficient proof that there is no substitute for monitoring Iraq's WMD program from inside the country, with the inspectors having the mandate to control anything relevant anywhere and anytime. In addition, the foreign inspectors' presence on the ground makes it much more difficult for Iraq to continue in its secret plans to develop WMD.[19] When former inspector Scott Ritter retreated from his UNSCOM post in August 1998, he declared that Iraq will be able to build an atomic bomb within six months after the end of ground inspections. Although Ritter later somewhat altered his position, there is some evidence that Saddam Hussein has revived his nuclear program following the ouster of the U.N. inspectors. Already in 1999, the International Atomic Energy Agency (IAEA) called it "essential" that inspectors return to Iraq because it is possible that Baghdad "retains the capability to exploit for nuclear-weapons purposes any relevant materials or technology to which it may gain access in the future, ... (and that Iraq retained) documents of its clandestine nuclear program, specimens of important components and possibly amounts of non-enriched uranium."[20] But what is true for the nuclear area is also valid for chemical and biological weapons as well as missiles: unless there is no monitoring inside Iraq, every attempt to assess the current state of Baghdad's WMD program cannot be more than "qualified (at best) guessing".

With regard to **nuclear weapons**, the greatest concern remains the possibility that, were Iraq to acquire fissile material from abroad, it could probably produce an operational nuclear weapon within a very short period of time. Whereas it is proven that Iraq obtained fissile uranium from South Africa until the end of the Apartheid regime, now sources from some successor states of the former Soviet Union are suspected to be involved in

[19] See Michael J. Eisenstadt, "Contain Broadly: Bolstering America's Current Iraq Policy", in Clawson, Patrick L. (ed.), *Iraq Strategy Review; Options for U.S. Policy*. Washington D.C.: The Washington Institute for Near East Policy 1998, p. 14.

[20] Quoted in James H. Noyes, "Fallacies, Smoke, and Pipe Dreams: Forcing Change in Iran and Iraq", in *Middle East Policy*, 7 (2000) 3, p. 32.

the Iraqi nuclear program. As in the case of drug smuggling, one can assume that for every disclosed case, lots of others go unreported.[21] Thus, it is not possible till the present day to state – without the slightest doubt – whether Iraq has nuclear weapons or not. Over the years, there was more than one report in the international media that tried to convince its readers of Saddam Hussein possessing the bomb. It was only in January 2001 that the *London Telegraph* published a story – allegedly based on a report by an Iraqi defector – that Iraq obtained "two fully operational weapons".[22] Media and defectors might have their own intentions, but a look to more serious sources as for instance major Western intelligence and defense institutions gives information that Iraq does not have enough enriched material for a nuclear weapon and that it will still last some years to obtain it.

Even President George W. Bush had to admit in his Cincinnati speech at 7 October 2002 that "we don't know exactly ... how close Saddam Hussein is to developing a nuclear weapon ... and that's the problem. Before the Gulf War, the best intelligence indicated that Iraq was eight to ten years away from developing a nuclear weapon. After the war, international inspectors learned that the regime has been much closer – the regime in Iraq would likely have possessed a nuclear weapon no later than 1993. The inspectors discovered that Iraq ... was pursuing several different methods of enriching uranium for a bomb ... (UNSCOM and IAEA) dismantled extensive nuclear weapons-related facilities ... (but) evidence indicates that Iraq is reconstituting its nuclear weapons program. Saddam Hussein has held numerous meetings with Iraqi nuclear scientists, a group he calls his 'nuclear mujahideen' ".[23] When talking of "different methods" and "numerous meetings", President Bush was probably referring to a report the BBC had televised in early March 2001 claiming that before the Second Gulf War Iraq had set up two groups, each pursuing different methods, to develop an atomic bomb, and that UNSCOM had failed to detect and destroy one of these, known as Group 4. The detailed report – once again based on information of an exiled Iraqi nuclear scientist – said that Group 4 had worked on a nuclear device very similar to the one used by the U.S.

---

[21] See Mark Gaffney, "Will the next Mideast War go nuclear?", in *Middle East Policy*, 8 (2001) 4, pp. 101 – 2.

[22] Jessica Barry, "Saddam has made two atomic bombs, says Iraqi defector", in *London Telegraph*, 28 January 2001.

[23] http://www.whitehouse.gov/news/releases/2002/10/print/20021007 – 8.html.

in Hiroshima. In 1998 the IAEA had reported that it had no evidence of an ongoing nuclear weapons program in Iraq but when former IAEA-chief and present UNMOVIC-head Hans Blix was asked by the BBC in March 2001, he had to confirm that also the IAEA knew nothing about the existence of Group 4 when it was still working inside Iraq.[24]

After having gassed thousands of innocent Kurds in Halabja in 1987, Baghdad could not deny having produced **chemical weapons** (CW). But the Iraqi government insisted on having terminated any production of CW after the First Gulf War. Nevertheless, when UNSCOM inspectors excavated missile warheads from sites where the Iraqi army had secretly destroyed weapons after the Second Gulf War, it became evident that they had once contained VX. Thus, the Iraqi asseverations of never having "weaponized" this lethal chemical were exposed as a lie. UNSCOM's findings were later confirmed by the defected former chief of Iraqi intelligence, General Wafiq as-Samarrai.[25] Thus, UNSCOM had many reasons to complain in its annual reports to the Security Council since 1991 of Iraq's unwillingness and/or inability to provide clear figures. For example, the 1998 report informed that Baghdad seized from inspectors an Air Force document discovered by UNSCOM indicating that Iraq had not consumed as many CW munitions during the Iran-Iraq War in the 1980s as had been declared by the Iraqi leadership. This discrepancy indicated that Iraq may have an additional 6.000 CW munitions hidden.[26]

The permanent "cat and mouse" game played by the Iraqi government illustrates the difficulty of UNSCOM's task. Nevertheless, the inspectors were able to trace stocks of blister and nerve agents, including quantities of "VX salt", a highly lethal nerve agent, more than 600 tons of VX precursors and some 4.000 tons of other precursor chemicals. In addition, they found between 30.000 and 40.000 munitions that could be filled with chemical or biological agents (including dozens of missile warheads, 2.000 bombs, 15.000 artillery shells, and 15.000 to 25.000 rockets).[27] The U.S. govern-

[24] See *The Middle East and North Africa, 2001*, Cambridge 2002, p. 447.

[25] See Andrew Cockburn, Patrick Cockburn, op. cit., pp. 111, 267.

[26] See Anthony H. Cordesman, "If we fight Iraq: Iraq and its Weapons of Mass Destruction", in *Gulfstates Newsletter*, 2 June 2002, p. 8.

[27] See Michael J. Eisenstadt, "Residual WMD Capabilities", in Clawson, Patrick L. (ed.), *Iraq Strategy Review; Options for U.S. Policy*. Washington D.C.: The Washington Institute for Near East Policy 1998, pp. 171–2.

ment's White Paper of 13 February 1998 revealed that Iraqi forces delivered additional CW including Mustard 5 agent and the nerve agents Sarin and Tabun in aerial bombs, aerial spray dispensers, 120 mm rockets, and different types of artillery. In spite of frequent destruction of Iraqi CW, UNSCOM was sure in early 1998 that Iraq still had a small stockpile of CW agents, munitions, and production equipment. Baghdad's report of having destroyed 200 metric tons of chemical precursors, 70 Scud warheads, and tens of thousands of smaller unfilled munitions was not verified.[28]

UNSCOM's intense activities until the end of that year, together with the precise bombardments of "Operation Desert Fox" in December let to the above mentioned assessment that "the disarmament phase of the Security Councils' requirements is possibly near its end in the missile and chemical weapons ... area", but it is also clear that Baghdad retained the expertise and the experience to quickly resume the production of CW. In the absence of any monitoring, Iraq could begin a small mustard agent production within a few weeks, a full-scale production of Sarin within a few months, whereas pre-Gulf War production levels of VX would take some years. Western intelligence revealed that the number of plants manufacturing chemicals had been greatly expanded since 1998, and that around 20 of them were directly linked to arms production.[29] Even if there is a high grade of guessing in the exact number of these production sites, it seems to be very likely that Iraq possesses CW at present.

Biological weapons (BW) are even harder to track, and it were only Hussein Kamil's revelations of 1995 that had delivered more than hints that such a program existed at all in Iraq. Kamil spoke not only of different secret sites producing BW, but also of 25 special warheads. According to his information, 16 of these warheads were filled with botulinum toxin, 5 with anthrax and 4 with aflatoxin. Under optimum conditions, the first two agents could kill more than a million people. On the targets he admitted that Israel and the Allied Forces as well as some Arab states who had supported "Operation Desert Storm" had been chosen. Immediate disclosures in Iraq revealed that the warheads were still existing. They were hidden in tunnels (railway or irrigation) or buried on the banks of the Tigris river to hide them from allied bombing raids. After a few months, the Iraqi site promised to

[28] See "Iraq Weapons of Mass Destruction Programs", op. cit., pp. 3–4.

[29] See *The Middle East and North Africa, 2001.*, op. cit.

have destroyed everything at a desert site called Nabai, but UNSCOM was never sure that this information was really true. According to experts there is more than a vague probability that the warheads are still hidden away somewhere in the deserts of Iraq.[30]

Nevertheless, after Hussein Kamil's revelations, Baghdad had to admit that Iraq had produced 8.500 liters of anthrax, 19.000 liters of botulinum toxin, and 2.200 liters of aflatoxin. Besides the above mentioned 25 warheads, Iraqi officials also admitted preparing BW-filled aerial bombs and aerial dispensers and acknowledged having tried to use 155 mm artillery shells, artillery rockets, a MIG-21 drone, and aerosol generators to deliver BW agents.[31] Although UNSCOM intensified its efforts after the revelations to destroy Iraq's BW, the effectiveness was questionable. It proved exceptionally difficult to find biological weapons, which require a relatively small infrastructure, and which can be easily concealed as ordinary medical or research facilities. Additionally, the Iraqi government did everything possible to prevent biowarfare inspections and remained especially intransigent – in comparison with nuclear and chemical weapons as well as long-range missiles – about what it had done with all its BW assets. According to experts, there were still about 30 tons of biological warfare agents unaccounted for when UNSCOM had to leave Iraq by the end of 1998; enough to poison – in theory – the entire world.[32] Among the BW were unknown quantities of seed stock and/or bulk stocks of anthrax, botulinum toxin, clostridium perfringens, aflatoxin, and ricin, seventeen tons of growth media to produce anthrax, equipment to be used for the production of biological agents in dried form which is much more effective when disseminating the agent than the liquid form which Iraq had acknowledged producing so far. UNSCOM experts also supposed that Iraq had possibly more advanced warhead designs than those recovered until 1998, and spray equipment that could be used to disseminate BW from manned or unmanned aircraft.[33]

---

[30] See Al J. Venter, "New-Era Threat: Iraq's Biological Weapons", in *Middle East Policy*, 6 (1999) 4, pp. 107–8.

[31] See "Iraq Weapons of Mass Destruction Programs", op. cit., p. 3.

[32] See Al J. Venter, op. cit., p. 105; President Bush stated in his Cincinnati speech that Iraq had probably produced four times more BW than its government had ever declared. See http://www.whitehouse.gov ... , op. cit.

[33] See Michael J. Eisenstadt, "Residual ... ", op. cit., p. 172.

The experts were also sure that Iraq could begin production of BW within a few days without effective U.N. monitoring. For instance, Iraq could easily convert the production of biopesticides to anthrax by simply changing seed material. In addition, anthrax can be stored and remains viable for many years. Only recently, U.S. intelligence analysts have been closely examining satellite images of the west bank of the Tigris River for signs of a large laboratory that should exist there – according to numerous rumors. The lab is purported to have 85 employees and the secret mission to produce BW.[34]

Before the Second Gulf War, Iraq had an active force of ballistic missiles that included operational Scud B missiles with a range of 300 km (imported from the Soviet Union), an advanced program to extend the Scud's range and improve its warhead (al-Hussein with a 650 km range and al-Abbas with a 950 km range) as well as to indigenously produce **long-range missiles** (Condor) that never left the design phase.[35] As mentioned above, UNSCOM was able to destroy – with minor uncertainties – Iraq's ballistic missiles. However, the Western intelligence community believes that Baghdad – although it is only permitted to produce missiles with a range of 150 km or less – has been able to maintain the infrastructure and expertise necessary to develop long-range missiles. This is a crucial point in general as David Kay, previously a leading nuclear weapons inspector on Iraq, warned. Even if UNSCOM would have destroyed all Iraqi WMD facilities, "the weapons secrets are ... well understood by a large stratum of Iraq's technical elite ... Iraq has now become more like post-Versailles Germany in its ability to maintain a weapons capability in the teeth of international inspections".[36] Therefore, Iraq's capabilities in this regard should not be underestimated.

For example, the poor showing of Scuds during the Second Gulf War could lead to the conclusion that as long as Iraq is limited to Scud-type missiles with a range of less than 150 km – and some believe that Baghdad has indeed only a small, covert force of those missiles, launchers, and Scud-specific production equipment and support apparatus – the real danger is very low. But what about Iraq's skilled technicians acquiring advanced

[34] See *Washington Post*, 30 July 2002.

[35] See "Iraq Weapons of Mass Destruction Programs", op. cit., p. 5.

[36] Quoted in Mark Strauss, "Attacking Iraq", in *Foreign Policy*, (2002) 129, p. 16.

gyroscopes and guidance systems from Russia or other previous Soviet republics as often suspected?[37] General Lee Butler, former commander of U.S. strategic forces gave a convincing answer in this regard: "There is really no piece of information, no piece of software or hardware, relating to ballistic missile technology that is not available to anyone who is willing to pay the price."[38]

The question remains open, however, whether the production of long-range missiles in Iraq – even if possible in its material and technological aspects – would go undetected by the sophisticated Western intelligence. In any case, President Bush was sure in September 2002 that "Iraq possesses ballistic missiles with a likely range of hundreds of miles – far enough to strike Saudi Arabia, Israel, Turkey, and other nations – in a region where more than 135.000 American civilians and service members live and work. We've also discovered through intelligence that Iraq has a growing fleet of manned and unmanned aerial vehicles that could be used to disperse chemical or biological weapons across broad areas".[39] The statement did not make a clear distinction between "ballistic" and "long-range" missiles and did not specify the "manned and unmanned aerial vehicles". If the latter is a prescription of the converted L-29 trainer aircraft which Iraq originally acquired from former Czechoslovakia than the danger seems to be low.

The U.S. National Intelligence Council, for instance, recently summarized the Iraqi ballistic missile threat to the U.S. as follows: "Although the Gulf war and subsequent United Nations activities destroyed much of Iraq's missile infrastructure, Iraq could test an Inter-Continental Ballistic Missile (ICBM) capable of reaching the United States during the next 15 years. After observing North Korean activities, Iraq most likely would pursue a three-stage Taepo Dong-2 approach to an ICBM ..., which could deliver a several-hundred kilogram payload to parts of the United States. If Iraq could buy a Taepo Dong-2 from North Korea, it could have a launch capability within months of the purchase; if it bought Taepo Dong engines, it could test an ICBM by the middle of the next decade. Iraq probably would take until the end of the next decade to develop the system domestically.

[37] See Vladimir Orlov and William C. Potter, "The Mystery of the Sunken Gyros", ***Bulletin of the Atomic Scientists***, 54 (1998) 6.

[38] Quoted in *Washington Post*, 29 July 1998.

[39] http://www.whitehouse.gov ... op. cit.

Although much less likely, most analysts believe that if Iraq were to begin development today, it could test a much less capable ICBM in a few years using Scud components and based on its prior SLV experience or on the Taepo Dong-1. If it could acquire No Dongs from North Korea, Iraq could test a more capable ICBM along the same lines within a few years of the No Dong acquisition. Analysts differ on the likely timing of Iraq's first flight test of an ICBM that could threaten the United States. Assessments include unlikely before 2015; and likely before 2015, possibly before 2010—foreign assistance would affect the capability and timing."[40] The time range, as well as the many "ifs" suggest – in connection with the UNSCOM findings and activities – that there is no immediate long-range missile threat from Iraq.

IV. Threat targets

There were many attempts to understand why the Baath leadership of Iraq is so determined to develop, produce and preserve weapons of mass destruction. In general, Washington has tended to explain this determination with the personal ambitions of Saddam Hussein. And indeed, the Iraqi president is driven by a dream to become a pan-Arab hero, the unifier of the Arab world, the avenger of the masses suppressed by Western imperialism etc. In this regard, it is the overall image that counts. Therefore, Saddam thinks that his open and secret supporters and admirers judge his performance by the ability to retain his WMD. In this sense, it is not the military value of the weapons alone that drives his actions. Would this be the case, the Iraqi president could easily have surrendered his entire WMD to UNSCOM inspectors right after the cease-fire in 1991, get a clearance bill and begin a clandestine rebuilding program immediately after the inspectors left Iraq. But for Saddam Hussein, the preservation of Iraq's WMD has always been a critical element of his power.[41]

In any way, Saddam's actions always deserve a look at the underlying motives. Yes, he is a brutal dictator who ignited two devastating full-scale wars within a decade, but he was and is not "crazy" or irrational. He may

[40] Quoted in Anthony H. Cordesman, op. cit.

[41] See Daniel Byman/Kenneth Pollack/Matthew Waxman, "Coercing Saddam Hussein: Lessons from the Past", in *Survival*, 40 (1998) 3, p. 141.

live in his own world screened by yes-sayers but it is stating the obvious to say that the Middle East is not like present-day Europe. It resembles far more the Europe of the 19th century that was characterized by the use of brute force between the major powers and by mutual fears of aggression. Therefore, the arms race between Middle Eastern countries has never ended, and the distrust between them remains strong. This situation led to a phenomenon which can be called "political hysteria" – a term introduced by the Hungarian philosopher and political scientist Istvan Bibo when analyzing the situation in Eastern Europe in the wake of the Second World War. Bibo wrote that – in contrast to developments in the West – "national identity in Eastern Europe was to be created, to be re-established, or to be obtained by fierce struggle. It had to be jealously preserved not only against the designs of dynastic states, but against the indifference of part of the population and against eclipses of national awareness ... (in this sense, 'political hysteria') was explained by a fear for the existence of the community which was further fired by the fragile character of the states in Eastern Europe".[42]

This analysis sounds familiar when applied to the Middle East, where the injuries stemming from colonialism are not forgotten, "where nationalisms are the more asserted the more uncertain they are; and where conspiracy theory and fear of one's neighbors replace all rational analysis. From this point of view, and even if the Iraqi leadership has developed to an excessive degree a paranoid vision of the world – divided between 'them' and 'us' – Iraq is not an isolated example of such behaviour ... Relations in this region are marked by the fact that each country fears its neighbours".[43]

Thus, several states in the region are attempting to acquire or develop WMD. These weapons are viewed – not only in Iraq – as symbols of prestige, means to counter superior conventional forces, and vehicles to spread influence. But in Iraq's case, history and experience suggest that the intention to possess and to use WMD is limited primarily to a national and secondarily to a regional dimension.

After the gassing of Kurds in Halabja, Saddam Hussein wants his people to be sure that he can use similar weapons against them all, anytime.

[42] Quoted in Alain Gresh, op. cit., p. 70.
[43] Ibid.

For example, in April 1999 the government deployed troops around the Shii holy city of Najaf who were wearing gas masks and special white overalls seemingly designed to protect them against chemical or biological weapons. The appearance of this special troops caused naked terror in Najaf where the Shiites had well-founded fears that the regime will use poison gas against them. Everybody stayed at home, the streets were deserted.[44]

The regional dimension has offensive as well as defensive aspects. In 1998/99, the Iraqi president provoked many heads of state in the region when he demanded that the Arabs should stand up and topple the corrupt "throne dwarves" who had alienated themselves from their people by deliberately collaborating with the United States.[45] Thus, Arab as well as Turkish leaders had enough reason to assume that Saddam Hussein would likely use a deliverable WMD capability to secure an Iraqi regional hegemony.

But on the other hand, Saddam himself has a strong feeling of being besieged by enemies. Therefore, he wants to expand his arsenal of WMD for defensive purposes, too. For instance, when Turkey and Iran supported the establishment of allied-controlled no-fly zones in northern and southern Iraq respectively, the Iraqi leadership was sure that both countries were looking for a partition of Iraq. Iraqi generals were sure that it was the use of chemical weapons and missiles that ultimately worked in favor of Iraq in the First Gulf War and that it were Iraq's WMD that had prevented the allied forces from conquering Baghdad in the Second Gulf War.[46]

This perception by far exceeds the person of Saddam Hussein. Most probably, any government in Baghdad will be eager to defend the national integrity of Iraq against threats from its enemies, no matter if they are real or imagined. Chas Freeman, president of the U.S. Middle East Policy Council, was sure that Iraq, like North Korea and Iran, is seeking to develop WMD, but was surprised why his compatriots think that changing Iraq's regime would alter this policy. "Isn't it possible that a democratically-elected government in Iraq might be interested in developing weapons of mass de-

---

[44] See Amatzia Baram, "The Effect of Iraqi Sanctions: Statistical Pitfalls and Responsibilities", in *Middle East Journal*, 54 (2000) 2, pp. 221–2.

[45] See Mark Strauss, op. cit., p. 15.

[46] Charles Duelfer, a former UNSCOM inspector, had extracted this impression from numerous discussions with Iraqi officials and officers. Hearing (U.S. Policy Towards Iraq) of the House International Relations Committee, Subcommittee on the Middle East and South Asia, Rep. Benjamin A. Gilman (NY), Chairman, 4 October 2001.

struction, just as a democratically elected Israel has done?"[47] he asked. If the use of WMD seems to be the last resort against an "Iranian threat", no Iraqi leader will voluntarily accept their abandonment. And, given the bombing of the Osirak reactor in 1981, how will Iraq deal with the "Israeli threat"?[48] Responsible and far-sighted Western policy has to take this reality into account – and "political hysteria" is part of that reality. Seth Carus, a proliferation expert at the National Defense University stated already in 1998 that "it certainly makes sense (for Arab states) to hold on to and continue to develop (WMD) programs as a deterrent to Israel's nuclear program".[49]

Israel has never signed the Nonproliferation Treaty, which requires every member country to allow inspections of its nuclear facilities, as well as the 1972 Biological and Toxin Weapons Convention, which was ratified – for example – by Iran, Libya and Saudi Arabia and signed by Syria and Egypt.[50] It is therefore undeniable that Israel's possession of WMD was and is generally stimulating proliferation in the region. It is the Western double standard in this regard that provides Iraq with a good excuse. Concerning WMD, the U.N. has inflicted a sanctions regime on Iraq, but indulged Israel. On invasions, Iraq has been sanctioned too, but there is remarkable silence with regard to the Chinese invasion of Tibet and the occupation/annexation of several of its neighbors by Israel.[51] Article 14 of UNSCR 687 demands to establish "in the Middle East a zone free of weapons of mass destruction and all missiles to deliver them, and the objective of a global ban on chemical weapons".[52] This – unfulfilled – demand made it easy for the Iraqi vice-president Ramadan to declare that Iraq is against the possession of WMD and that the Security Council should meet its obligations in accordance with its own resolutions pertaining to Iraq and implement paragraph

[47] Quoted in http://www.arabnews.com/Sarticle.asp?ID=18509&set=U.S.%20conference.

[48] See Alain Gresh, op. cit., p. 72.

[49] Quoted in *Washington Post*, 14 April 1998.

[50] See James H. Noyes, op. cit., p. 32.

[51] See Thomas Stauffer in "The end of Dual Containment: Iraq, Iran and Smart Sanctions", in *Middle East Policy*, 8 (2001) 3, pp. 84–5.

[52] Quoted in Phyllis Bennis/Stephen Zunes/Martha Honey, "The Failure of U.S. Policy Toward Iraq and Proposed Alternatives", Ibid., p. 102.

14 of resolution 687, which is related to dismantling WMD in the whole region, particularly the "weapons of the usurping Zionist entity".[53]

Knowing all this, it sounds very simplistic how General Tommy Franks, Commander-in-Chief of U.S. Central Command (CENTCOM), explained the Iraqi threat: "Iraq's ambition is to assume leadership of the Pan-Arab world and, as such, Iraq represents a continuing threat to short-term regional stability. Its attempted annexation of Kuwait 12 years ago and continued efforts to subvert U.N. sanctions through economic blackmail are a continuing challenge to our interests. Repeated attempts to down coalition aircraft, continued refusal to accept the U.N. weapon inspections regime, and threatening gestures toward Iraq's neighbors indicate Saddam Hussein has not been deterred from his desire to dominate the region."[54] At least with regard to the region he is right, but – according to all accounts – Iraq, as a state actor, definitely lacks the means to threaten the West with WMD. But the questions remains whether this is also true with regard to non-state actors, i.e. terrorists who might get support from Iraq.

V. Terrorism

Without doubt, Saddam Hussein has a record of organizing and supporting terrorism. Who will count the numerous victims of his brutal terror against alleged or real opponents within Iraq. Between 1998 and 1999 for example, he killed all of the most senior leaders of the majority Shia community of Iraq, including Grand Ayatollah Muhammad Sadiq as-Sadr and his sons, and the Ayatollahs Borujerdi and Gharavi.[55]

On the regional level, terrorist activities he supported reach from car bombs in Damascus to assassinations in Amman, Abu Dhabi or Kuwait, where an attempt to kill the Emir and his guest, former U.S. president Bush, failed in 1993.[56] Furthermore, the Baath regime shelters several terrorist groups: among them are the Mujahedin-e Khalq (MEK), the Kurdistan Workers' Party (PKK), the Palestine Liberation Front (PLF), and the Abu

[53] Iraqi TV, 13 February 2002, 18:00 GMT, Quoted in *BBC Monitoring Global Newsline*, 14 February 2002.

[54] *Gulfwire e-Newsletters*, 28 February 2002, p.4.

[55] Patrick L. Clawson, "Rethinking Iraq Strategy: Why and How?", in Clawson, Patrick L. (ed.), op. cit., p. 1.

[56] See *The Observer*, 16 December 2001.

Nidal Organization (ANO) that allegedly carried out – until the recent death of its founder in Baghdad – more than 90 terrorist attacks in two dozen countries which killed or injured almost 900 people, including 12 Americans.[57] Giving shelter to these organizations manifests indirect terrorism at best. But in October 1998, Iraqi agents were said to have planned attacking the U.S.-funded Radio Free Europe/Radio Liberty (RFE/RL) station in Prague, which produces programs in Arabic for Iraqi listeners and employs journalists who are declared enemies of Saddam Hussein. Although the attack never happened, the mere plan seems to prove the assumption that the main target of terrorism carried out by Iraq itself is on dissident Iraqi activity in the West. So what about any international, particularly Western or American targets of direct or indirect Iraqi terrorism?

The Iraqi government was the only Arab leadership that did not condemn the 11 September terrorist acts against the United States declaring that America was now "reaping the fruits of its crimes against humanity".[58] Furthermore, Saddam Hussein's eldest son Udayy has said that the 11 September incidents in the U.S. were "daring operations which were carried out by Arab Muslim youths and which have caused the U.S. to take heed of Arabs and Muslims among other peoples".[59] On that basis some U.S. officials were quick at hand to say that there "is growing evidence of contacts between Iraq and Osama bin Ladin's Al-Qaeda network".[60] During that time, the accusations were mainly based on the findings of Iraq-watcher Laurie Mylroie who had focused her research on possible connections between bin Ladin's Al-Qaeda and Sudanese intelligence that had invited the Saudi islamist to Khartoum in 1991. Given the close ties between Iraq and Sudan (Khartoum supported Baghdad during the First Gulf War and became subsequently a major base for Iraqi intelligence), she was asking if Al-Qaeda was in contact with Iraqi intelligence in Sudan, and if Abd al-Samad al-Taish, the highest ranking Iraqi agent in Khartoum between

[57] See http://www.whitehouse.gov . . . , op. cit.

[58] Quoted in *Patterns of Global Terrorism – 2001,* Washington D.C.: U.S. Department of State 2002, Iraq.

[59] Quoted in *Ash-Sharq al- awsat,* 10 February 2002.

[60] *Financial Times,* 4 November 2001.

1991 and 1998, ever met Osama bin Ladin there.[61] Mylroie's answer can only be summarized in a "possibly".

Other U.S. analysts were positive that Iraq sent agents – at least twice – to meet bin Ladin or his aides when he went back to Afghanistan in 1996. They claimed that senior Iraqi intelligence official Farouk Hijazi visited Kandahar in December 1998, a few months after the bombings in Kenya and Tanzania; however, it was not clear whether he came in contact with bin Ladin.[62] Thus, since academic and intelligence analysts could not present "water-proof" evidence of direct links between Iraq and Al-Qaeda, the media took over. Newspapers claimed that Saddam's officials had frequently met with Al-Qaeda representatives handing out $ 3.8 million in funding altogether, and that bin Ladin was using Iraq's banking network.[63] If that would be really true, it could be worth searching for a direct link between the 11 September attacks and orders from Iraq.

But although some senior U.S. intelligence sources declared to have evidence that hijackers Marwan ash-Shehi and Ziad Jarrah were contacted by Iraqi agents, the former CIA director James Woolsey who is not hiding his deep personal commitment to topple Saddam Hussein, had to admit that he was unable to prove a link between Iraq and the 11 September attacks. Authorities in Prague vehemently rejected a report stating that the leading terrorist pilot Muhammad Atta had met with an Iraqi agent in the Czech capital some months before 11 September. In the meantime, it is also clear – based upon intense laboratory tests – that the anthrax sent by mail throughout the United States in the fall of 2001 was not produced in Saddam's weapons labs but originated in the U.S.[64]

At present, after tremendous efforts by intelligence analysts as well as academic and political experts, not a single piece of evidence was found linking Iraq itself or via Al-Qaeda with a terrorist attack against the U.S.

---

[61] See *Wall Street Journal*, 13 September 2001. Laurie Mylroie is the author of *Study of Revenge: Saddam Hussein's Unfinished War Against America*, Washington D.C.: American Enterprise Institute 2000. In the book Mylroie cites the late FBI's New York director James Fox who was sure that Iraq provided support, probably even coordination, to the terrorists who bombed the World Trade Center in 1993. The author reports of three other attacks against the U.S. with a possible connection to Iraq.

[62] See "Iraq", *Gulf States Newsletter (GSN) Extra*, 17 October 2001.

[63] See for example *The Sunday Times*, 16 September 2001.

[64] See Mark Strauss, op. cit., p. 16.

or American citizens after the failed attempt against ex-president Bush in 1993. Therefore, as inhuman and brutal proven Iraqi terrorism against dissidents might be, it cannot be used by the U.S. as a casus belli. On the contrary, there is no reason to assume that Saddam Hussein may start terrorist acts against the U.S. now, when he refrained from doing so during the last decade. Only when Washington, from its part, will drive Saddam into a corner he may choose terrorism as a last resort.

VI. Threat assessment

In the face of a looming war against Iraq, it is of extreme importance to be right in assessing the threat constituted by present-day Iraq and its leadership. Unfortunately, most of the statements, reports and analyses seem to be severely influenced by political interests. There is obviously one strong faction in the U.S. administration, including the president, that ascribe to Iraq a clear and present danger.

As President George W. Bush declared in Cincinnati: "Eleven years ago, as a condition for ending the Persian Gulf War, the Iraqi regime was required to destroy its weapons of mass destruction, to cease all development of such weapons, and to stop all support for terrorist groups. The Iraqi regime has violated all of those obligations. It possesses and produces chemical and biological weapons. It is seeking nuclear weapons. It has given shelter and support to terrorism, and practices terror against its own people ... the threat from Iraq stands alone – because it gathers the most serious dangers of our age in one place. Iraq's weapons of mass destruction are controlled by a murderous tyrant who has already used chemical weapons to kill thousands of people ... If the Iraqi regime is able to produce, buy, or steal an amount of highly enriched uranium a little larger than a single softball, it could have a nuclear weapon in less than a year ... And Saddam Hussein would be in a position to pass nuclear technology to terrorists."[65] Some might argue that there is more assertion in this speech than hard evidence, and that it can be doubted why Saddam should pass nuclear weapons to terrorists when he did not do it with his deployable CW and BW, but – nevertheless – the president's speech is a mirror of a widespread opinion; also based on the assessment of high-ranking militaries.

[65] http://www.whitehouse.gov ... , op. cit.

For instance, on 27 February 2002, CENTCOM-chief General Tommy Franks, declared when addressing the House Services Committee: "The three years absence of the U.N. arms monitors from Iraq has permitted Baghdad to pursue ballistic missiles and weapons of mass destruction. This absence, and unmonitored cross-border traffic, have provided opportunities to acquire sensitive and dual-use materials. We assess that Iraq continues to pursue biological, chemical, and prohibited ballistic missile capabilities."[66]

But there are also academics who support this political-military approach. According to Anthony Cordesman, Iraq has rebuilt key portions of its chemical production infrastructure for industrial and commercial use, as well as its missile production facilities since the end of the Second Gulf War. The government has attempted to purchase numerous dual-use items for, or under the pretext of, legitimate civilian use. This equipment – in principle subject to U.N. control – also could be diverted for WMD purposes. Of course, since the suspension of U.N. inspections in December 1998, the risk of diversion has increased. Cordesman continued that Baghdad instituted a reconstruction effort on those facilities destroyed by "Operation Desert Fox" in December 1998, to include several critical missile production complexes and former dual-use CW production facilities. In addition, Iraqi specialists appear to be installing or repairing dual-use equipment at CW-related facilities. Some of these facilities could be converted fairly quickly for production of CW agents.[67]

Quite naturally, also the exiled Iraqi opposition is supporting this approach. In an effort to get Saddam in the American line of fire, Sheik Muhammad Ali, a co-founder of the INC, declared that "the regime of Saddam Hussein is connected with all terrorist groups in the world and I think that he supported the hijackers in the September 11 attacks and had connections with Muhammad Atta ... Also, the regime is still producing weapons of mass destruction, including biological and chemical weapons, and I think this will lead to more terrorist attacks throughout the world ... the United States has a responsibility to support the Iraqi people and the Iraqi opposition in toppling this regime ... (which) will be much easier than the current campaign in Afghanistan".[68] Other opposition forces in-

[66] *Gulfwire e-Newsletters*, 28 February 2002, p.8.

[67] See Anthony H. Cordesman, op. cit.

[68] http//www.Kurdishmedia.com, 9 February 2002.

side Iraq claim that Saddam Hussein is hiding vast stocks of WMD in the myriad of newly-built mosques and even in the compounds of hospitals.[69] Nevertheless, nothing is sure in this regard, and in the absence of proven facts, any attempt to take own accusations as the truth could lead to hazardous consequences. And there were always more sober assessments of the Iraqi threat.

According to leading experts, UNSCOM activities have uncovered and destroyed much of Iraq's WMD capabilities. In a lecture at Chatham House, Fred Halliday spoke of 95%.[70] Even former UNSCOM official Scott Ritter, an outspoken falcon, concluded that monitoring "allowed UNSCOM to ascertain, with a high level of confidence, that Iraq was not rebuilding its prohibited weapons programs and that it lacked the means to do so without an infusion of advanced technology and a significant investment of time and money".[71] This statement coincides with a report by the IAEA that was declaring – even before "Operation Desert Fox" – that Iraq was free of nuclear weapons and the capability to produce them.[72] Last but not least, at the same time Hussein Shahristani, former chief scientific adviser to the Iraqi energy program, insisted at a conference in Cambridge (U.K.) that Iraq's nuclear capacity has been largely liquidated. Nevertheless, Shahristani added that he was not equally sure with regard to CW and BW.[73] But even in this case, experts are not certain whether Iraqi officers are really able to master essential technologies. This is especially referring to the fuse technology necessary to detonate chemical warheads. Obviously, the successful delivery of a chemical agent is a complicated matter. A chemical warhead will only function properly, if it dispenses the liquid agent in an aerosol form in the moment prior to impact.[74]

But not only technological aspects count. In a wider dimension, sanctions have stripped Baghdad of the military, political and economic influence it had before the Second Gulf War. The embargo has prevented Saddam Hussein from acquiring materials to fully restore his military-industrial

[69] *Tariq ash-Sha'b*, 14 March 2002.

[70] Cited in Said K. Aburish, *Saddam Hussein: The Politics of Revenge*, London: Bloomsbury 2000, p. 360.

[71] Quoted in Walid Khadduri, op. cit., pp. 158 – 9.

[72] See *International Herald Tribune*, 21 April 1998.

[73] See *The Middle East and North Africa, 2001.*, op. cit.

[74] See Mark Gaffney, op. cit., p. 103.

base and has – in spite of smuggling – severely limited clandestine arms imports, while the general atmosphere of frustration caused by the sanctions has caused a deep demoralization of the armed forces.[75]

The dove faction is not only composed of academics and former UNSCOM inspectors but of politicians, too. For example, the former U.N. Assistant Secretary General and Humanitarian Coordinator for Iraq, Graf Sponeck, was very clear when stating: "Intelligence analysis has not detected convincingly any links between Iraq and international terrorism nor has it confirmed an existing capacity of Iraq to produce weapons of mass destruction. The recent decision by the U.K. government not to publish a report claiming to show that Iraq remained in possession of WMD is further evidence to this effect. It also enhances the significance of the statement by former U.S. Secretary of Defense William Cohen when he briefed incoming President Bush on January 10, 2001, that "Iraq no longer constituted a military threat to its neighbors".[76]

Obviously, there is an interrelationship between the impossibility to paint a clear picture of Iraq's present WMD status and the wide contradictions in the threat assessments.

VII. Conclusions

The hitherto presented information leads to three interrelated personal conclusions:

A) The danger of Saddam Hussein using his arsenal of WMD – how operable it ever might be – has seldom be as low as it is at present. Everything the U.N. is demanding now from Saddam Hussein – not to speak about American demands – is worse than the existing status quo for him. Therefore, Saddam Hussein is mainly interested in the stabilization of the present situation. This is even true with regard to the sanctions regime. The embargo has helped him to foster his grip on power, and to blame the U.N. and the West for the misery of the Iraqi population. And then there is the lucrative smuggling.

Since the beginning of the embargo, Iraq operates an oil pipeline from Kirkuk to Ceyhan in Turkey, that is capable of transporting about 900.000

---

[75] See Patrick L. Clawson, op. cit., p. 2.

[76] Quoted in *Jordan Times*, 12 July 2002.

barrels of oil per day (bpd) and is running at near full capacity. Nevertheless, the oil-for-food program closely regulates output from this pipeline. However, truck tankers smuggle oil from Iraq to Turkey. According to the U.S. Energy Information Administration, Iraq also smuggles oil to Iran across the Fao Peninsula with barges. In addition, there have always been reports that Iraq smuggles oil by truck to the Mediterranean via Syria and Lebanon. But until 2000, none of Iraq's existing truck or barge smuggling routes was capable of moving the 200.000 – 400.000 bpd of oil that has been smuggled out of the gulf every day. In March 1999 however, Iraq's deputy oil minister declared that the pipeline from Kirkuk in northern Iraq to the Syrian Mediterranean port of Banias that had been shut off since the First Gulf War was repaired. Syria agreed to reopen the pipeline capable of transporting about 300.000 – 350.000 bpd. The bilateral agreement became effective in November 2000. Altogether, Iraq earns approximately $ 3 billion per year on the black market by selling oil and fuel at discounts up to 40% below market prices to neighboring countries like Syria, Turkey, and Jordan. The money is flowing directly into government coffers evading any international influence or control.[77]

For sure, in the those countries, Saddam Hussein is still not be seen as the worst choice. On the contrary, Baghdad has become, in the last few years, the main trading partner and aid donor to countries like Jordan, Syria and Lebanon through oil grants. Iraq has also signed free trade agreements with many countries in the region (including Egypt), promising to become a key partner to any state that challenges American policies. Most of the neighboring states are afraid of the alternatives and the consequences of a regime change in Iraq. Turkey fears any development that might make Iraq's Kurds more independent. Syria is concerned that a pro-American regime in Baghdad might encircle it. Riyadh fears the Lebanonization of Iraq or the southern parts floating towards Iran. Jordan fears a radical Islamist take over in Baghdad and/or a flood of tens of thousands of Iraqi refugees. Iran prefers a weakened Saddam to a new regime that would be – almost inevitably – aligned with Washington. After Hamid Karzai was in-

[77] See Toby Dodge, "Dangerous Dead Ends?", in *The World Today*, 57 (2001) 7, p. 8.

stalled in Kabul, the most important of Iran's neighbors would then be U.S.-allies.[78]

B) All the previous experiences with Saddam Hussein suggest that he is more interested in the here and now than in the future. He refrained from using chemical or biological weapons against allied forces during the Second Gulf War, since he was aware of a threat from Washington that he personally and Iraq would "pay a terrible price."[79] And indeed, the defected Iraqi WMD program director Hussein Kamil declared in August 1995 in Amman, that the Iraqi leadership refrained from using chemical weapons out of fear that the U.S. might respond with tactical nukes.[80] Also later on, Saddam Hussein from time to time (obviously in 1993) stopped to interfere with UNSCOM in an attempt to improve relations with the United States. When there were signals that he intended to conquer Kuwait for a second time in 1994, he gave in when facing a massive American military buildup.

C) Saddam was never close enough to being toppled during the 1991 Gulf War to use his weapons of mass destruction as a last resort. But he is thoroughly prone to doomsday thinking. During the Second Gulf War he ordered the highest-ranking commanders of his missile forces to attack Israel with WMD if communications between them and Baghdad were interrupted and if the commanders got the impression that Baghdad had fallen to the Allied forces. This order could have unleashed an unconventional war, since Saddam Hussein had to expect the likelihood of a nuclear response by Israel. Therefore, the order meant nothing else than Saddam's preference of Baghdad being annihilated rather than conquered by the Allied forces.[81] It would be extremely hazardous to hope that at present – in a similar case – a given commander of the missile force would not carry out such orders. Even President Bush was reflecting this concern – from another angle however – in his Cincinnati speech: "An Iraqi regime faced with its own demise may attempt cruel and desperate measures. If Saddam Hussein orders such measures, his generals would be well advised to refuse those orders. If they

[78] See Ghassan Atiyyah, "Dealing with Saddam or his clone", in *Middle East Policy* 7 (2000) 4, p. 145.
[79] See Mark Strauss, op. cit., p. 15.
[80] See Mark Gaffney, op. cit., p. 103.
[81] See also Amatzia Baram, op. cit., p. 14.

do not refuse, they must understand that all war criminals will be pursued and punished."[82]

But even in the case of the majority of Iraqi soldiers defecting after the beginning of an U.S. strike against Iraq or most of the generals refusing to carry out orders, Saddam will be left with a maneuvering room of at least three to four days. He has learned his lessons from the 1991 Gulf War and changed – for example – his communications technology. The new, subterranean cables are made of silk-fiber, delivered by North Korea, which is not detectable from the air. Three or four days are enough time to inflict heavy damages on enemies especially when using WMD. And unlike 1991, Saddam would be sure this time that his survival is at stake, and would use his entire arsenal.[83]

Therefore, Iraq's neighbors are deeply concerned. Will the United States really liberate the Iraqi people from its dictator while simultaneously protecting them? If this is not secured, if Washington will leave Saddam in power with all his WMD, then the neighbors are afraid of his revenge. A half-hearted American initiative might only lead to an enraged Iraqi president who could look for relief by attacking his neighbors. Furthermore, they are not convinced of the seriousness of the American war plans at all. "Our problem is that we see much of it as wishful thinking or a leap of faith – particularly relying on defections. This doesn't have the feel of a workable plan," said an Arab official. "None of us are defending Saddam Hussein," he added, "but we want to make sure that everyone is better off the day he is gone, and that means a lot more planning than is going on now."[84] And by the way, these are the most optimistic Arab voices.

Thus, to resume in one sentence: There is not enough evidence of the imminent danger of Saddam Hussein's WMD program to risk a military campaign that could cause more harm and instability for the peoples of the region as well as for the whole world than the *potential* threat the Iraqi Baath regime is constituting might do in the future.

[82] http://www.whitehouse.gov . . . , op. cit.

[83] Interview of the author with a representative of the Patriotic Union of Kurdistan (PU.K.) at 18 February 2002 in Damascus.

[84] Quoted in Robin Wright, "Bush's Team targets Hussein", in *Los Angeles Times*, 10 February 2002.

Sanctions, Inspections and Containment. Viable Policy Options in Iraq*

David Cortright, Alistair Millar, George A. Lopez

I. Introduction

Concerns are growing that Iraq may be rebuilding its capacity to develop and use weapons of mass destruction. After more than three years without U.N. inspections, the uncertainties and risks associated with Iraq's weapons programs have increased. The urgency of these issues has prompted widespread calls for the resumption of U.N. weapons inspections, and has led U.S. officials to threaten military attack. The U.S. threats are also motivated by a desire to overthrow the government of Iraq. Pundits in the United States have raised a chorus of calls for military action to topple Saddam Hussein.

Many leaders in the region support the goal of disarming Iraq, but as U.S. vice president Dick Cheney learned during his March trip to the Middle East, most of these same leaders oppose U.S. military action against Iraq. States in the region fear the consequences of a U.S.-led war, especially in light of the profound security crisis in the Middle East.[1] These

---

* First published in: Policy Brief Series F3, June 2002. Reprinted with permission of the Joan B. Kroc Institute for International Peace Studies, University of Notre Dame, Indiana. All rights reserved.

[1] U.N. Secretary-General Kofi Annan has spoken for many world leaders in raising critical concerns about U.S. military action in Iraq. See "UN Chief Warns against Iraq Attack", *BBC News Online,* 10 December 2001, available at <http://news.bbc.co.uNotesk/hi/english/world/middle_east/newsid_1701000/1701002.stm> (11 April 2002); Barbara Crossette, "Annan Plans Talks with Iraq on Inspections Plan", *New York Times,* 26 February 2002, A6; and "UN Chief: The US will not act Wisely if it Attacks Iraq", *ArabicNews.Com,* 26 February, 2002, available at

realities suggest the need for viable alternative strategies to resolve the Iraq crisis and protect regional security.

This report presents policy options available to the United States for addressing security concerns in Iraq. It examines the issues associated with the threat of weapons development in the region and offers a series of policy options for reducing and containing that threat without resort to military force. The report does not dwell on the uncertainties and risks of waging war on Iraq without international consent. These have been amply examined in other articles and commentaries.[2] The paper concentrates instead on robust alternatives to the use of force. The policy options outlined here include:

- Reforming U.N. sanctions to tighten controls on oil revenues and military-related goods while further easing restrictions on civilian economic activity;
- Facilitating the return of U.N. weapons inspectors to complete the U.N. disarmament mandate and reestablish an Ongoing Monitoring and Verification (OMV) system; and
- Creating an "enhanced containment" system of externally based border monitoring and control if Iraq refuses to allow the resumption of weapons inspections.

The report begins with an assessment of Iraq's capacity for developing weapons of mass destruction. It then examines options for controlling Iraq's weapons potential through economic statecraft, United Nations weapons inspections, and diplomatic engagement with neighboring countries.

<http://www.arabicnews.com/ansub/Daily/Day/020226/2002022613.html> (11 April 2002).

[2] Among the recent reports and articles raising questions and concerns about war in Iraq are the following: Philip H. Gordon and Michael E. O'Hanlon, *Should the War on Terrorism Target Iraq?*, Policy Brief 93 (Washington, D.C.: The Brookings Institution, January 2002); Michael W. Isherwood, "US Strategic Options for Iraq: Easier Said than Done", *Washington Quarterly* 25, no. 2 (Spring 2002): 145–59; Mark Strauss, "Think Again: Attacking Iraq", *Foreign Policy* (March/April 2002); Ivo H. Daalder and James M. Lindsay, "Next Stop: Iraq", *San Jose Mercury News*, 2 December 2001; Vincent M. Cannistraro, "Keep the Focus on Al Qaeda", *New York Times*, 3 December 2001, A23; Stephen Fidler and Roula Khalaf, "Back to Iraq", *Financial Times*, 1 December 2001, 8; Jessica Mathews, "The Wrong Target", *Washington Post*, 4 March 2002.

II. The Nature of the Threat

There is no doubt that the regime of Saddam Hussein poses a significant threat to regional and international security. The regime has initiated two wars and has developed and used chemical weapons and ballistic missiles against neighboring states and its own citizens. Baghdad's 1988 attacks against Halabja and other Kurdish villages serve as a grim reminder of the regime's readiness to use the most horrific instruments of mass murder.[3] In the aftermath of the Gulf War, U.N. officials discovered that Iraq was acquiring the ability to develop nuclear weapons and had vast stockpiles of chemical and biological weapons. Some experts estimated at the time that Iraq was only a year or two away from producing a deployable nuclear weapon.[4]

As a result of the destruction caused by the Gulf War and the extensive weapons monitoring and dismantlement efforts of the United Nations Special Commission (UNSCOM), much of Iraq's capacity for developing and using weapons of mass destruction was eliminated during the 1990s. Since the departure of UNSCOM from Iraq in 1998, however, weapons monitoring and dismantlement efforts have come to a halt. U.N. officials have been unable to determine the status of Iraq's weapons programs for more than three years, although U.S. and other intelligence services have continued to gather information on suspected weapons activities. Iraqi defectors have also provided assessments of Iraqi weapons programs, although these reports are difficult to verify.[5] In light of these reports and Iraq's past behavior, it is prudent to assume that the Baghdad government has been attempting to rebuild its weapons capacity.

Nuclear weapons pose the greatest danger, but they also require the greatest effort to develop. Estimates of Iraq's present capabilities in this area vary, but expert testimony suggests that Baghdad is still several years away from achieving nuclear weapons status. A January 2001 report from the U.S. Department of Defense noted that "Iraq would need five or more years

[3] See the gripping account of the 1988 chemical weapons attacks against Kurdish villages in Jeffrey Goldberg, "The Great Terror", *The New Yorker*, 25 March 2002, 52–75.

[4] "The World's Nuclear Arsenal", *BBC News Online*, 2 May 2000, 14:49 GMT, at <http://news.bbc.co.uk/hi/ english/world/newsid_733000/733162.stm> (2 May 2000).

[5] For a recent example, see David Rose, "Iraq's Arsenal of Terror", Vanity Fair (May 2002): 120–31.

and key foreign assistance to rebuild the infrastructure to enrich enough material for a nuclear weapon".[6] Former U.S. assistant secretary of state for nonproliferation Robert Einhorn estimated in the summer of 2001 that Iraq was five years away from being able to produce a nuclear explosive.[7] August Hanning, the chief of the German Intelligence Agency (BND) was more pessimistic in a recent interview with *New Yorker* writer Jeffrey Goldberg: "It is our estimate that Iraq will have an atomic bomb in three years."[8]

U.S. and U.K. officials claim to have evidence that Iraq is developing prohibited ballistic missile technology.[9] U.S. officials recently showed Security Council members satellite photographs and documents that reportedly provide evidence of an Iraqi project to build prohibited long-range missiles.[10] Some analysts contend that Iraq retains a small force of Scud-derived missiles, and that work is proceeding on the *Al Samoud* liquid-propellant missile.[11] Experts report that Iraq has also attempted to extend the range of the short-range missiles it was permitted to retain after the Gulf War, although so far without success. In the area of chemical and biological weapons, it is likely that Iraq retained some stockpiles of chemical weapons after UNSCOM's departure, and that it also possesses considerable biological weapons potential. The latter category is the least amenable to control and elimination, due to the dual-use nature of many biological ingredients and precursor elements.

Hans Blix, the head of the United Nations Monitoring, Verification and Inspection Commission (UNMOVIC), which is charged with weapons in-

---

[6] Department of Defense, Office of the Secretary of Defense, Proliferation: Threat and Response (Washington, D.C.: U.S. Government Printing Office, January 2001), 40.

[7] Robert Einhorn, "The Emerging Bush Administration Approach to Addressing Iraq's WMD and Missile Programs", keynote address at the conference, "Understanding the Lessons of Nuclear Inspections and Monitoring in Iraq: A Ten-year Review", June 2001; sponsored by the Institute for Science and International Security, Washington, D.C.; available at Institute for International Security <http://www.isis-online.org/publications/Iraq/Einhorn.html> (15 April 2002).

[8] Goldberg, "The Great Terror", 75.

[9] Colum Lynch, "Mideast Crisis Delays Campaign at UN to Expose Alleged Threats to Obtain Prohibited Weapons", Washington Post, 7 April 2002, A22.

[10] William Ofre, "US says Iraq is Developing Banned Arms Weapons", Los Angeles Times, 4 May 2002, available at <http://www.latimes.com/news/printededition/asection/la-000031593May04.story?co> (4 May 2002).

[11] Charles A. Duelfer, "Viewpoint", Aviation Week and Space Technology, 11 March 2002, 73.

spection within Iraq, told a reporter recently that he has seen satellite imagery of new construction at possible weapons sites in Iraq and has received tips about other potential weapons activities. Blix noted that nothing has been proven, however, and he emphasized that evidence of such activity must be brought to the U.N. Security Council.[12] The legal authority for addressing the Iraqi weapons threat belongs to the Council, not any single government.

Although the potential threat posed by Iraq is considerable, it is important not to exaggerate the regime's military capacity. The combined results of war, more than a decade of stringent sanctions, and the previous weapons dismantlement efforts of UNSCOM have significantly diminished the Iraqi military threat. According to reports by UNSCOM and the International Atomic Energy Agency (IAEA), U.N. weapons inspections effectively neutralized much of Iraq's ability to develop and use weapons of mass destruction. The independent panel of experts established by the Security Council in 1999 concluded, "In spite of well-known difficult circumstances, UNSCOM and the IAEA have been effective in uncovering and destroying many elements of Iraq's proscribed weapons programs ... The bulk of Iraq's proscribed weapons programmes has been eliminated."[13] The IAEA stated in 1998 that "there is no indication that Iraq possesses nuclear weapons or any meaningful amounts of weapon-useable nuclear material".[14] Of the 819 *Scud* missiles known to have existed at the start of the Gulf War, UNSCOM accounted for all but two of these missiles. UNSCOM found no evidence of successful indigenous missile development, and no indications of prohibited missile testing. UNSCOM also reported "significant progress" in destroying chemical weapons stockpiles and production facilities. U.N. inspectors were less successful in eliminating biological weapons capabilities.

Despite Iraq's repeated efforts to deceive and disrupt U.N. weapons inspectors, UNSCOM succeeded in dismantling a considerable part of Iraq's

[12] Lynch, "Mideast Crisis".

[13] United Nations Security Council, *Letter Dated 27 March 1999, from the Chairman of the Panels Established Pursuant to the Note by the President of the Security Council of 30 January 1999 (S/1999/100) Addressed to the President of the Security Council*, S/1999/356, New York, 30 March 1999, par. 25.

[14] Quoted in United Nations Security Council, *Letter Dated 27 March 1999*, S/1999/356, par. 14.

most threatening weapons programs and capabilities. Although uncertainties remain, especially regarding biological and chemical weapons, the most dangerous threats—nuclear weapons and long-range missiles—have been substantially reduced. The lack of a credible ballistic missile capacity is especially significant in limiting the regime's ability to deliver chemical or biological weapons against neighboring states or the military forces of other nations. The Iraqi air force has less than half the strength it possessed at the time of the Gulf War (when it was no match for U.S.-led forces) and has only minimal capacity to deliver any type of weapon of mass destruction or to threaten neighboring states. Its ill-equipped bomber force is estimated to consist of just six planes.[15] The possibility of Iraq using a single plane or missile to attack Israel cannot be discounted, of course, but it is extremely unlikely that Saddam Hussein would commit such an act, unless he is faced with a large-scale military attack designed to overthrow his regime.

The threat that Iraq or any other state poses is a function of both capability and intention. These are mutually reinforcing factors, but they are also distinct and can be assessed separately. The current wave of excitement in Washington about attacking Iraq has blurred these distinctions. Some analysts mistakenly assume that Iraq's intentions translate into military capability, while others assume that any weapons activity, even that permitted under Security Council resolutions, implies aggressive intention. Pundits and some officials in Washington assume that, based on Saddam Hussein's past actions and intentions, it is only a matter of time before Iraq develops weapons of mass destruction and uses them against either Israel or the United States. But no expert can pretend to know with certainty the personal goals and policy aspirations that govern the actions of Saddam Hussein and his government. Even if these assumptions are correct, the logical emphasis of U.S. policy should be denying the regime the means of realizing these intentions. Currently there is no U.S. plan short of war for achieving this objective. It is the very prospect of war, however, that is most likely to motivate the regime to use whatever weapons capability it may possess.

---

[15] Air Force figures based on Anthony Cordesman, *The Military Balance in the Middle East; The Northern Gulf: Iraq* (Washington, D.C.: Center for Strategic and International Studies, 28 December 1998), 32 and 64, available at <http://www.csis.org/mideast/reports/mbmeXiraq122898.pdf> (12 April 2002).

The continuing U.N. sanctions against Iraq have hampered the regime's ability to rebuild its weapons capacity. Although sanctions have not been successful in convincing the Baghdad government to comply with U.N. mandates, they have been effective as a means of military containment. Sanctions have prevented the Baghdad government from gaining access to its vast oil revenues. The U.N., not the Baghdad government, controls most of the income derived from Iraqi oil sales. Since the beginning of sanctions, it is estimated that the Baghdad government has been denied more than $150 billion in oil revenues.[16] As a result, Iraq has been unable to purchase sufficient weapons and military-related goods to rebuild and modernize its armed forces. The cumulative arms import deficit for Iraq since 1990 is more than $50 billion.[17] This is the amount of money Iraq would have spent on weapons imports if it had continued to purchase arms as it did during the 1980s. Although Iraq gains some unrestricted revenue through smuggling and kickbacks (estimated at between $1.5 and $3 billion annually),[18] this income is not sufficient to fund a large-scale military development program. As a result, Iraq's ability to produce weapons of mass destruction and the means to deliver them has been curtailed. The picture that emerges from this assessment, then, is of a regime committed to the redevelopment of weapons of mass destruction, but constrained by diminished resources and the successes of the previous UNSCOM weapons dismantlement effort. The evidence indicates that Baghdad has not yet fully reconstituted its nuclear weapons and ballistic missile capabilities, and that it may be several years away from doing so. There is still time to thwart Iraq's acquisition of weapons material and technology, to constrain the regime's ability to finance weapons purchases, and to resolve the current crisis through economic and diplomatic means. Failing that, the United States and the United Nations should act to strengthen the military containment of the regime.

[16] See Meghan O'Sullivan, *Iraq, Time for a Modified Approach*, Policy Brief 71 (Washington, D.C.: The Brookings Institution, February 2001), 4.

[17] Based on estimates by Anthony Cordesman, *Military Expenditures and Arms Transfers in the Middle East* (Washington, D.C.: Center for Strategic and International Studies, July 2001), 79, available at <http://www.csis.org/burke/mb/milexpenditurearmstransfer.pdf> (17 August 2001).

[18] Estimates based on Raad Alkadari, "The Iraqi Klondike", *Middle East Report* (Fall 2001); Carola Hoyos, "Oil Smugglers Keep Cash Flowing", *Financial Times*, 17 January 2002; and Alex M. Freedman and Steve Stecklow, "How Saddam Diverts Millions Meant for Food Aid to Reap Illegal Oil Profit", *Wall Street Journal*, 2 May 2002.

The policies and mechanisms sketched below constitute a more realistic strategy for achieving these objectives than the pursuit of armed regime change.

III. Reforming Sanctions

In May 2002 the Security Council adopted Resolution 1409, fundamentally reshaping U.N. sanctions in Iraq. Under the terms of the resolution, restrictions on shipping civilian goods to Iraq were lifted. The arms embargo remained in place, and a new technology transfer control system was established. The focus of sanctions thus shifted from restricting civilian trade to prohibiting the import of weapons and military-related goods. The resolution approved a Goods Review List (GRL) of specific dual-use items that would be subject to review and approval, as outlined in an annex of procedures attached to the new resolution. The review procedures would apply only to designated dual-use technologies and goods with potential weapons application. All other civilian goods would be permitted to flow freely into Iraq without monitoring or preapproval.

The adoption of Resolution 1409 was the culmination of more than a year of deliberation at the Council. It reflected a desire by Council members to provide humanitarian relief for Iraqi civilians, and to shift the burden of responsibility for any further social hardships from the Security Council to the Baghdad government. The new policy also demonstrated the Council's commitment to maintaining more targeted, focused pressure on Iraq's weapons programs. Resolution 1409 created a more sustainable U.N. policy of sanctioning weapons and military-related technology.

Resolution 1409 was introduced by the five permanent members of the Council and was approved by a unanimous vote of all fifteen members, including Syria. This unity within the Council reflected the emergence of a new consensus on U.N. policy in Iraq. Despite Saddam Hussein's attempts to undermine U.N. sanctions and splinter the coalition arrayed against him, the U.N. Security Council has become more united on the need to retain controls on military-related imports to Iraq.

Political dynamics in the Council have improved considerably in the last two years, as indicated in the most important Iraq-related resolutions. In December 1999, Russia, France, and China abstained on the vote approving Resolution 1284, which created UNMOVIC. At that time the permanent members were deeply split and unable to agree on a formula for resuming weapons inspections. Since then the political climate for constraining Iraq's ambitions has improved. When the Council considered a new U.K./U.S. sanctions proposal in June 2001, France and China indicated support for the measure and rejoined the majority. Only Russia objected. When the Council adopted Resolution 1382 in November 2001 initially approving the GRL, Russia dropped its objections. The consensus deepened in the spring of 2002 when the United States and Russia approved the list of specific dual-use items that would be subject to GRL review.[19] This agreement on issues that directly affect the Russian economy was of considerable significance, given Moscow's long-standing ties to Baghdad's military and oil industries. It would be a setback for U.S. policy in Iraq, and for U.S. relations with Russia, to forego this cooperation in favor of unilateral military action. Despite widespread "sanctions fatigue" and sharp differences among the major powers over policy in Iraq, the Security Council has managed to maintain political unity on reforming sanctions.

As the Security Council considers additional reform measures, its goal should be to retain those elements of sanctions that have been effective—restrictions on military-related goods—while removing those elements, such as limitations on civilian trade, that have caused humanitarian hardships. The Council should also adopt measures to prevent the smuggling of oil. The Security Council should consider the following additional options for sanctions reform:

Establish Better Control of Oil Pricing and Marketing.

Kickbacks and illegal payments to Iraq have been continuing problems with the sanctions regime. To discourage Baghdad from requiring such payments, and to make it more difficult for oil purchasers to provide them, members of the Iraq sanctions committee began to introduce "retroactive pricing" in the fall of 2001. U.S. and British members of the committee delayed the approval of the official selling price of Iraqi oil.[20] The so-called price adjustment period was reduced from thirty to fifteen days. This is the

[19] Peter Baker, "US and Russia Agree to Overhaul Sanctions in Iraq", *Washington Post*, 29 March 2002, A20.

[20] Freedman and Stecklow, "How Saddam Diverts Millions".

time between contract approval and the actual delivery and purchase of the oil. The purpose of this change was to prevent Iraq from providing substantial discounts to buyers in return for back-door payments. The new system of retroactive pricing succeeded in making such payments more difficult. Buyers were no longer able to receive what a leading U.N. official called "an abnormally high premium for Iraqi crude oil".[21]

As with so many aspects of the Iraq sanctions regime, however, the change in the oil pricing mechanism had unintended negative consequences. It led to a reduction in oil exports and a consequent decline in the revenues available for the purchase of humanitarian goods. In early 2002 the export rate of Iraqi oil dropped to 1.4 million barrels a day, well below the previous rate of more than 2 million barrels a day.[22] Oil purchasers who had profited from discounted oil were no longer "satisfied with more reasonable premia", according to the director of the U.N. Iraq Programme. The buyers refused to sign new contracts and postponed or cancelled previous ones.[23]

The Security Council found itself caught in a dilemma. The introduction of retroactive pricing made it more difficult to cheat on the sanctions, but it also led to a drop in the revenues available for the humanitarian program. Members of the Security Council have considered a number of new proposals for adjusting the pricing mechanism so that traders can be sure that the value of their contract matches the price they must pay on delivery. The Council and the U.N. Office of the Iraq Programme have also sought ways to monitor and control the companies and trading organizations authorized to purchase and deliver Iraqi oil, as further means of preventing illegal payments. These efforts should continue as the Council seeks to balance the need for adequate revenues for civilian imports and the need for effective controls against kickbacks to Saddam Hussein.

---

[21] United Nations, Office of the Iraq Programme, "Statement by Benon V. Sevan, Executive Director of the Iraq Programme at the Informal Consultations of the Security Council", New York, 26 February 2002, 4.

[22] "Statement by Benon V. Sevan", 3.

[23] "Statement by Benon V. Sevan", 4.

Permit Controlled Foreign Investment.

Although the so-called oil for food program authorized under Resolution 986 (1995) has steadily expanded over the years, permitting a wide range of imports covering oil production, telecommunications, transportation, and many other civilian sectors, a further easing of trade restrictions is warranted. Additional measures to facilitate and encourage investment and civilian trade will not only help to ease the humanitarian hardships caused by sanctions, but will create new economic opportunities within Iraq and among trading partners.

Some limited and controlled forms of foreign investment in Iraq should be allowed to facilitate industrial development and speed economic recovery. Investments in the oil sector would enable Iraq to increase its production capacity, thereby generating additional revenues for the purchase of nonmilitary supplies. A special committee of monitors should be established to set criteria for foreign investment and to recommend ways of preventing Iraq from using investment income for military purposes.

IV. Facilitating the Return of Weapons Inspectors

The most effective way to prevent the Baghdad regime from redeveloping weapons of mass destruction is to return U.N. weapons inspectors to the country. The priority for U.S. policy must be to convince the Iraqi government to permit the resumption and completion of the U.N. disarmament mandate. As noted, previous U.N. weapons monitoring efforts achieved significant progress in ridding Iraq of weapons of mass destruction. Its successor agency, UNMOVIC, is trained and ready to carry on the task of completing the disarmament effort. It is vital that the United States take advantage of the renewed consensus in the Security Council to give UNMOVIC the opportunity to perform its duties. Effective U.N. weapons inspections offer the best hope for detecting and destroying Iraq's weapons of mass destruction.

U.N. and U.S. officials have correctly insisted on the right of free and unfettered access for UNMOVIC inspectors. The monitors must have access to so-called presidential sites and other suspected weapons locations. Although it is appropriate to insist on unrestricted access once on site, it is also important to be flexible in negotiating the return of inspectors so that they can begin their work. Absolute access to every square inch of Iraqi territory is neither possible nor necessary. Some degree of Iraqi participation in the process of determining access is inevitable, and should be accepted as the price of returning monitors to the country. Once the inspectors are back in Iraq, they may gain access to unanticipated information and evidence. At a minimum, inspectors would be able to reestablish an Ongoing Monitoring and Verification (OMV) system, as was previously installed by UNSCOM. Ideally this would include the right of unannounced inspections at undeclared, sensitive locations. Among other things the OMV system would permit the monitoring of Iraq's airways and waterways, to test for radioactive particles that would indicate the presence of nuclear weapons activity. The benefits of restoring monitoring access far outweigh the risks of accepting some ambiguities in UNMOVIC's terms of reference.

Some observers are concerned that the United States may prevent the resumption of inspections by demanding unreasonable standards of access for UNMOVIC. It is no secret that some hardliners in.**7** June 2002 Washington dismiss the reestablishing of inspections as a diplomatic trap, and an impediment to the preferred use of military force.[24] U.S. Secretary of Defense Donald Rumsfeld recently expressed open skepticism that U.N. weapons inspectors would be able to detect Iraqi weapons capabilities.[25] This is an unrealistic and shortsighted view that ignores the significant success of previous weapons dismantlement efforts. It was UNSCOM, not U.S. bombing that achieved the greatest progress in reducing the Iraqi weapons threat. It is UNMOVIC, not a new war that can resolve the remaining weapons issues with the greatest degree of international cooperation and support.

In creating UNMOVIC the Security Council set out a timeline of approximately one year for the completion of the U.N. weapons monitoring and dismantlement mission.[26] Meeting this timeline will depend on Iraqi

[24] Former UNSCOM inspector Charles Duelfer wrote in January 2002 that negotiations over the return of U.N. inspectors pose "big problems" for the United States. "Any system Iraq would accept is not likely to be intrusive enough", complained Duelfer. See Charles Duelfer, "Inspectors in Iraq? Be Careful What You Ask For", *Washington Post*, 9 January 2002, A19.

[25] Robert Burns, "Rumsfeld, in break with State Department, says Iraq would deceive and deny U.N. weapons inspectors, *AP,* 16 April 2002.

[26] "UN Inspector Tells Council Work in Iraq Could be Fast", *New York Times*, 22 March 2002, A7.

government cooperation. It will also depend on providing concrete assurances to Iraq that cooperation with UNMOVIC will bring benefits in the form of relief from more than a decade of sanctions.

The diplomatic key to persuading Iraq to accept renewed weapons inspections is an effective carrots-and-sticks bargaining strategy. This requires both coercion and persuasion. Iraqi fears of a possible U.S. military attack are a coercive factor that may influence the regime's willingness to cooperate. A more widely supported form of coercive pressure is the U.N.'s continuing control over Iraqi oil revenues. Despite smuggling and kickback schemes, the U.N. still controls more than 80 percent of Iraq's oil income.[27] The Baghdad government urgently wishes to regain control of these revenues, which at a production rate of 2 million barrels a day comes to nearly $20 billion a year. U.N. and U.S. officials can turn this Iraqi objective to their advantage by structuring an inducement plan that offers the prize of oil revenues in exchange for full compliance with U.N. mandates.

The proposed bargaining strategy carries some risks, but these are manageable through effective diplomacy. Inducement policies can create political and moral problems if they reward wrongdoing or give the appearance of appeasement. Making offers to aggressors can be seen as a sign of weakness and may embolden an outlaw regime to further acts of belligerence. Because of these concerns, any inducements offered to Iraq must be strictly conditional, with conciliatory gestures linked to clear and unequivocal concessions from the Baghdad regime. There can be neither concessions to intransigence nor any backing away from satisfactory completion of the U.N. disarmament mandate. The lifting of sanctions must be strictly conditioned on the certification by UNMOVIC and the IAEA that Iraq's capabilities for developing weapons of mass destruction have been fully eliminated.

[27] Total Iraqi oil sales in 2000 were nearly $18 billion. See basic figures provided by the United Nations Office of the Iraq Programme, available at *United Nations* <http://www.un.org/depts/oip/latest/basicfigures.html> (5 September 2001). Alkadiri estimated Baghdad's earnings from oil smuggling at no more than $1.5 billion annually, or approximately 8.3 percent of total revenues. If one accepts the larger estimates of smuggling revenue provided by diplomats, $3 billion annually, the rate rises to approximately 17 percent. *Wall Street Journal* reporters Freedman and Stecklow quoted U.S. State Department claims of $2.5 billion a year in illicit Iraqi oil revenue. Based on the estimates of Alkadiri, "The Iraqi Klondike"; Hoyos, "Oil Smugglers"; and Freedman and Stecklow, "How Saddam Diverts Millions".

The question of inducements for cooperation lies at the core of the Iraq impasse and is crucial to the challenge of finding a diplomatic solution. A clear and unequivocal commitment to lift sanctions and revenue controls upon fulfillment of the U.N. disarmament mandate could provide the necessary incentive to gain Iraqi cooperation. The United States has refused to consider any easing of coercive pressure, however, and has become fixated on the goal of armed regime change. The U.S. policy of unyielding hostility toward the Baghdad government has become a major obstacle to the resolution of the crisis.[28] Part of the strategy available to the U.S. is returning to the terms of the original Gulf War cease-fire Resolution 687 (1991), which specified in paragraph 22 that sanctions against Iraq will be lifted upon completion of the U.N. disarmament mandate. A clarification of this original Security Council obligation could help to gain Iraqi compliance.

A restatement of the Council's original intent would remove ambiguities left by Resolution 1284 (1999), which called merely for the suspension of sanctions rather than their termination. This weakened the previous unequivocal commitment to lift sanctions once the disarmament process is complete. Resolution 1284 also added a requirement for the affirmative renewal of the sanctions suspension every 120 days. This would allow the United States or another permanent member to use its veto power to halt the suspension and thus reimpose sanctions.

The Russian Federation offered a draft Security Council resolution in June 2001 that sought to clarify these ambiguities and that reaffirmed the obligation to lift sanctions upon completion of the U.N. disarmament mandate. Under the terms of the Russian proposal, once UNMOVIC and IAEA certified that a reinforced OMV system was fully operational within Iraq, sanctions would be suspended and oil revenues returned to the Iraqi government. The Russian proposal called for the continuation of a comprehensive arms embargo on Iraq. It also specified that the Security Council could terminate the suspension of sanctions upon evidence of Iraq acquiring prohibited military-related goods.[29]

---

[28] See our previous discussion of these issues in David Cortright and George A. Lopez, "The Limits of Coercion", *The Bulletin of the Atomic Scientists* 56, no. 6 (November/December 2002): 18–20.

[29] Permanent Mission of the Russian Federation to the United Nations, "Draft Resolution on Iraq", 26 June 2001.

Former UNSCOM inspector Garry Dillon proposed a similar approach at a June 2001 conference in Washington sponsored by the Institute for Science and International Security.[30] Under Dillon's plan, the Security Council would issue a new resolution lifting the oil embargo upon receipt of satisfactory assurances from UNMOVIC and the IAEA of Iraq's disarmament. The proposed resolution would reaffirm Iraq's obligation to permit the continued operation of the OMV system. It would also maintain the arms embargo and the provisions of Resolution 687 that prohibit Iraq from developing weapons of mass destruction.[31]

Paragraph 14 of Resolution 687 (1991) described the mandated disarmament of Iraq as a step toward "establishing in the Middle East a zone free from weapons of mass destruction". The Security Council thereby recognized the connection between Iraq's weapons programs and those of neighboring states in the Gulf Region and the Middle East. It is likely that any government in Iraq, either the present regime or a successor, will be motivated by balance of power considerations to match the capabilities of neighboring states. This suggests that the disarmament of one will only last if it is matched by the disarmament of all. The United States and other major powers must therefore work with the states of the region, including Israel, to seek the mutual elimination of weapons of mass destruction and a reduction of offensive military capabilities.

V. Creating an "Enhanced Containment" Border Monitoring System

If the Baghdad government does not permit resumed U.N. weapons inspections, it will be necessary to create an externally based, vigorously enforced system of enhanced military containment to restrict the flow of weaponsrelated goods into Iraq. The goal of the proposed system would be to establish a long-term capability for blocking Iraqi rearmament through strict controls

[30] Garry Dillon, "The Iraq Dilemma", paper presented at the 2001 Carnegie Nonproliferation Conference, June 2001, available at *Carnegie Endowment for International Peace* <http://www.ceip.org/files/projects/npp/ resources/Conference%202001/panels/dillon.htm> (15 July 2001).

[31] David Albright and Kevin O'Neill, "The Iraqi Maze: Searching for a Way Out", in *The Nonproliferation Review* (Fall/Winter 2001): 10.

on the import of weapons and dual-use military goods. Planning for the creation of such a system should begin now, in parallel with efforts to reform sanctions and encourage the reentry of weapons inspectors. A visible and credible effort to prepare for a sustainable system of enhanced containment might help to convince the Iraqi regime to accept the option of complying with U.N. weapons inspections.

An effective, externally based military control regime would depend on a multiple set of financial and technical restrictions and significant political and diplomatic initiatives to gain the cooperation of the states neighboring Iraq. Under the proposed system, current financial controls would remain in place. Iraq would not regain access to its oil revenues until it complied with Resolutions 687 and 1284 and allowed the resumption and completion of U.N. weapons inspections. The U.N. escrow account would be retained, and all purchases of unapproved imports would continue to require sanctions committee authorization. The retention of financial controls would preserve current restrictions on Iraq's ability to purchase military-related goods and weapons of mass destruction.

An enhanced military containment system would also require a significant strengthening of border monitoring in Jordan, Syria, Turkey, and other states surrounding Iraq. At present there is no international monitoring of the commercial crossings into Iraq. Shippers of approved humanitarian goods stop at the border to have documents authenticated, so that they can receive payment from the U.N. escrow account, but their cargoes are not inspected. The neighboring states have customs and border monitoring stations (and they gain revenues from duties on goods entering Iraq), but these controls are not designed to impede the flow of weapons.

The development of an enhanced military containment system would require the deployment of an adequately funded, well-equipped, and professionally trained international inspection force to detect and prevent shipments of nuclear materials or other prohibited items. To date the neighboring states have not supported proposals for border monitoring, in part because they do not want to disturb the growing commercial trade with Iraq that has developed in recent years. The challenge for the Security Council and U.S. policy is to design and create an effective system for inspecting sensitive cargoes, while avoiding disruption to the thriving civilian commerce that is vitally important to local economies.

Advanced monitoring and scanning technology can assist in the creation of such a border monitoring system. With appropriate equipment and resources, trained monitors should be able to detect the shipment of nuclear materials and other prohibited weapons-related goods without major disruption to commercial traffic.

The model for such a system might be the "smart border" program now being established by the United States, Canada, and Mexico. This program utilizes x ray-scanning equipment that can quickly inspect trucks and containers for contraband.[32] The equipment can safely and nonintrusively inspect containers at the rate of one per minute.[33] This would enable each equipment station to scan more than 700 trucks or containers in a 12-hour period. The "smart border" system also features an electronic pass system. Approved traders could be issued a machine readable electronic pass enabling them to cross the border quickly without inspection. Such passes could be issued to humanitarian agencies and other trusted suppliers of civilian goods financed through the U.N. Iraq account. Vehicles or containers with electronic passes would proceed without stopping; others would be required to pass through the X-ray detection equipment.

These technologies could be combined with customs support stations in which U.N.-approved international monitoring experts work alongside officials from the host nations to maintain and operate the detection equipment. This is the model of the successful Sanctions Assistance Missions (SAMs) that were developed for the U.N. sanctions in Yugoslavia during the years 1993 through 1995.[34] The assistance missions would not only help with the operation of advanced detection equipment, but could also provide general assistance in upgrading and improving border monitoring capabilities in the host countries.

---

[32] See Elisabeth Bumiller, "White House Announces Security Pact with Mexico", *New York Times*, 22 March 2002, A18.

[33] "New Non-intrusive Gamma Ray Technology to Scan Containers Through Port of Vancouver", press release, Vancouver Port Authority, 25 January 2002, available at *Yahoo Financial News* <http://biz.yahoo.com/ prnews/020125/va247_1.html> (26 January 2002).

[34] For a report on the Sanctions Assistance Missions, see United Nations Security Council, *Letter Dated 24 September 1996 from the Chairman of the Security Council Committee Established Pursuant to Resolution 724 (1991) Concerning Yugoslavia, Addressed to the President of the Security Council, Report of the Copenhagen Roundtable on United Nations Sanctions in the Case of the Former Yugoslavia, Held at Copenhagen on 24 and 25 June 1996*, S/1996/776, New York, 24 September 1996.

The task of monitoring shipments into Iraq would be a substantial challenge, but it would be less formidable than inspecting the large volume of traffic that crosses the U.S.-Mexican border every day, or that arrives in a busy port like Vancouver. Tens of millions of dollars of detection equipment and hundreds of trained professionals would be needed to operate the proposed border monitoring system, but these requirements would pale in comparison with those of a large-scale military operation. With appropriate technical capabilities and financial resources, a relatively nonintrusive but effective border control system in the countries surrounding Iraq could be created. Such a system would enable the Security Council to establish an externally based mechanism for enhancing the effectiveness of military sanctions. When combined with continued revenue controls, the proposed border control system could preserve military containment and help to prevent the redevelopment of weapons of mass destruction. No monitoring program can eliminate smuggling completely, but the proposed system could make illegal arms shipments more difficult and costly than they are now and could serve as a deterrent against smuggling.

Creating such a system would require a major commitment of financial and political capital. The economic costs of the proposed system could be charged to the U.N. escrow account, as part of the budget for U.N. operations in Iraq. Substantial financial support and technical assistance to frontline states would help to offset the costs of monitoring equipment and additional customs staffing, and would enable these governments to upgrade border control facilities and systems.

The greatest obstacles to creating an effective border monitoring system would be political, not financial or technical. An enhanced containment system depends on persuading frontline states to cooperate with the proposed monitoring mechanisms. This will involve extensive negotiations with Jordan, Syria, and other frontline states. The United States and other major powers must also be ready to offer substantial economic incentives and political assurances to these states. Iraq can be counted on to do everything in its power to undermine the proposed containment system. It will use its economic ties with neighboring states as leverage to threaten a cutoff of trade and oil supplies. Such pressure could have a devastating impact, especially in Jordan and Syria. The United States and other major powers must anticipate and counteract these pressures. They must be prepared to outbid Iraq by providing assurances of economic assistance and political support in the event of a cutoff of oil supplies and trade. This should be possible economically. Even with all of Iraq's oil wealth, the United States and other major powers can easily match the resources of Saddam Hussein. Saudi Arabia and other oil producers can compensate for the loss of Iraq's two million barrels a day.

The question is not resources but political will. Is the United States willing to open political relations with countries in the region previously considered inimical? Washington is now faced with taking bold diplomatic steps to achieve the containment of Saddam Hussein. Large-scale incentives and assurances enabled the United States to gain the support of Pakistan and several central Asian republics for its military campaign to overthrow the Taliban regime and disrupt al Qaida operations in Afghanistan. Similar steps are needed with Syria and Iran. Taking Syria off the list of states supporting terrorism would be a powerful inducement for gaining Syrian cooperation, which would be critical for controlling oil exports and limiting illegal payments. Iran and the United States cooperated in the initial stages of the campaign in Afghanistan, and each would benefit from continued cooperation to achieve the military containment of Iraq, among other mutual interests.[35] Even before 11 September, the United States was beginning to establish a more cooperative relationship with Sudan and Yemen, and was attempting to reassess its relations with Syria, although so far with little success. By building upon political openings with these and other countries in the region, the United States could forge vital political partnerships for a concerted diplomatic effort and cooperative border monitoring system to prevent Iraq from developing weapons of mass destruction.

Political cooperation with Russia would also be crucial to the proposed system. Moscow's support is essential if military containment is to be successful. Russia is the largest potential source of materials and technologies that could be used for Iraq's weapons of mass destruction. It has been implicated in past weapons smuggling incidents.[36][36] But Russia has supported a

[35] See Lee H. Hamilton, James Schlesinger, and Brent Scowcroft, "Thinking Beyond the Stalemate in US-Iranian Relations", *Policy Review* (Washington, D.C.: The Atlantic Council, May 2001), available at *Atlantic Council of the United States* <http://www.acus.org/> (9 October 2001).

[36] In 1995 Russian missile guidance components destined for Iraq were intercepted at the Amman airport, while other shipments of missile components were fished out of the

continuing arms embargo against Iraq in its proposals to the Security Council, and it has more to gain from cooperation with the West than from its ties to Saddam Hussein. Moscow and Washington have resolved many of their differences over sanctions reform and are cooperating across a broad range of international security issues. NATO is about to build an historic partnership with Russia for joint policies on counterterrorism, nonproliferation, and arms control. With imaginative diplomacy it should be possible to build upon this emerging pattern of synchrony to forge a joint approach to the containment of Iraq.

VI. Conclusion

The proposals outlined in this report present viable strategies for the continued denial of Iraq's weapons ambitions, while offering the prospect of greater regional cooperation and stability. They offer realistic alternatives to the military scenarios being discussed in Washington. In light of the dangers and uncertainties associated with what could be a large-scale and destructive war in the region, the nonmilitary options outlined here deserve immediate and thorough consideration. None of the proposed measures would be sufficient alone to achieve U.N. and U.S. objectives in Iraq, but taken together they offer an array of options for advancing international objectives without the risks of war. New U.N. action to reformulate sanctions, intensive diplomatic efforts to resume weapons inspections, and the creation of an enhanced containment system through revenue controls and strengthened border monitoring—these are the elements of the viable diplomatic alternative to war. They offer the best hope for meeting U.S. foreign policy objectives and enhancing security in the region.

Acronyms

BND	Bundesnachrichtendienst: German Intelligence Agency
GRL	Goods Review List

Tigris River near Baghdad. See "Russia's Dangerous Exports", Carnegie Endowment for International Peace, Washington, D.C., Non-Proliferation Fact Sheet, 19 August 1998, *at Carnegie Endowment for International Peace* <http://www.ceip.org/programs/npp/ factsheet2.htm> (18 April 2002).

IAEA	International Atomic Energy Agency
NATO	North Atlantic Treaty Organization
omv	Ongoing Monitoring and Verification System
SAMs	Sanctions Assistance Missions
UNMOVIC	United Nations Monitoring, Verification and Inspection Commission
unscom	United Nations Special Commission.

Authors

Gerhard Beestermöller, Dr. theol. habil., is Deputy Director of the Catholic Institute for Theology and Peace, Barsbüttel near Hamburg. His focus of research is Political Ethics and Peace Ethics. He recently authored: *Krieg gegen den Irak – Rückkehr in die Anarchie der Staatenwelt?* (Beiträge zur Friedensethik 35), Stuttgart: Kohlhammer, 2002; (Ed.) *Politik der Versöhnung* (Theologie und Frieden vol. 23), Stuttgart: Kohlhammer, 2002; (Ed.) *Die humanitäre Intervention – Imperativ der Menschenrechte?* (Theologie und Frieden vol. 24), Stuttgart: Kohlhammer, 2002.

Drew Christiansen S.J. is Assciate Editor of the America magazine and Counselor for International Affairs of the U.S. Conference of Catholic Bishops. Under the latest of his numerous articles are: *Hawks, Doves and Pope John Paul II*, in America, August 8, 2002, and *After September 11: Catholic Teaching on Peace and War*, in Origins, May 30, 2002.

David Cortright is President of the Fourth Freedom Forum, and a Research Fellow at the Joan B. Kroc Institute for International Peace Studies at the University of Notre Dame. Cortright was Executive Director of SANE, the largest U.S. peace organization, from 1977 to 1987. He has authored several books and numerous articles about peace issues. His most recent volumes are: *Smart Sanctions: Targeting Economic Statecraft* (Rowman & Littlefield, 2002, edited with George A. Lopez), and *Sanctions and the Search for Security: Challenges to UN Action* (Lynne Rienner Publishers, Inc., 2002, with George A. Lopez and Linda Gerber).

Klaus Dicke is Professor of Political Theory and History of Ideas at the Friedrich-Schiller-University in Jena. His research concentrates on United Nations, International Law and Peace Ethics, Human Rights, Modern Constitutionalism and more. Recent publications are: *Globales Recht ohne Weltherrschaft. Der Sicherheitsrat der Vereinten Nationen als Weltge-*

setzgeber?, Jena 2002; *Republik und Weltbürgerrecht* (co-ed. with Klaus-Michael Kodalle), Köln-Weimar 1998.

Henner Fürtig is Senior Researcher at the German Oriental Institute in Hamburg. His focus of research is contemporary history and politics of the Middle East with a special interest in the Gulf region. Recent publications are: *Iran's Rivalry with Saudi Arabia between the Gulf Wars*, Reading: Ithaca Press, 2002; *Islamische Welt und Globalisierung: Aneignung, Abgrenzung, Gegenentwürfe*, Würzburg: Ergon, 2001; *Iraq as a golem. Identity Crisis of a Westerns Creation*, in: Hafez, K. (ed.), *The Islamic Worl and the West. An Introduction to Political Cultures and International Relations*, Leiden et al: Brill, 2000, pp. 204–216.

Hans J. Giessmann is Professor of Political Sciences at the University of Hamburg and Deputy Director of the Institute for Peace Research and Security Policy (Hamburg). He works especially on the following topics: International Security Policy, Terrorism and on regional issues of Eastern Europe and Asia. From his most recent writings stand out: *Security Handbook 2001* (co-ed. with Gustav E. Gustenau), Baden-Baden: Nomos, 2001; *Multilateral Regional Security – OSCE Experiences and Lessons*, Seoul: Friedrich-Ebert-Stiftung, 2001.

John Langan S.J. is Joseph Cardinal Bernardin Professor of Catholic Social Thought, School of Foreign Service and Kennedy Institute of Ethics, Georgetown University, Washington. Areas of Specialization are: Ethics and Political Philosophy; his research interests include: ethics and international affairs, especially applications of just war theory; human rights in theory and practice; Catholic social teaching; the ethical theories of St. Thomas Aquinas and St. Augustine. He is currently working on a manuscript on the ethics of humanitarian intervention.

David Little is T. J. Dermot Dunphy Professor of Religion, Ethnicity and International Conflict and Director of the Center for the Study of Values in Public Life at Harvard Divinity School, Cambridge (MA). He publishes on the Ukraine and Sri Lanka.

George A. Lopez is Director of Policy Studies and Senior Fellow at the Joan B. Kroc Institute for International Peace Studies at the University of Notre Dame. Lopez focuses on economic sanctions and state violence and repression. His work on these subjects has been published in *Chitty's Law Journal, Human Rights Quarterly, The Bulletin of the Atomic Scientists,*

International Studies Quarterly, International Journal of Human Rights, Ethics and International Affairs, and *Fletcher Forum* as well as numerous books in which he has been author and editor.

Alistair Millar is Vice President of the Fourth Freedom Forum and Director of its Washington, D.C. office. Millar was a Senior Analyst at the British American Security Information Council. He has written on a wide range of issues, including sanctions, incentives, and nuclear nonproliferation. His opinion editorials and articles have appeared in several publications and periodicals including the *San Francisco Chronicle, Los Angeles Times, Defense News,* and the *Journal of International Affairs.*